COMPUTER SCIENCE, TECHNOLOGY AND APPLICATIONS

DEEP LEARNING AND ITS APPLICATIONS

Computer Science, Technology and Applications

Additional books and e-books in this series can be found on Nova's website under the Series tab.

COMPUTER SCIENCE, TECHNOLOGY AND APPLICATIONS

DEEP LEARNING AND ITS APPLICATIONS

ARVIND KUMAR TIWARI
EDITOR

nova
science publishers
New York

DOI: https://doi.org/10.52305/BLXA7086

NOTICE TO THE READER

Library of Congress Cataloging-in-Publication Data

ISBN: 978-1-68507-185-1

Published by Nova Science Publishers, Inc. † New York

CONTENTS

PREFACE

In just the past five years, deep learning has taken the world by surprise, driving rapid progress in fields as diverse as computer vision, natural language processing, automatic speech recognition, etc. This book presents an introduction to deep learning and various applications of deep learning such as recommendation systems, text recognition, diabetic retinopathy prediction of breast cancer, prediction of epilepsy, sentiment, fake news detection, software defect prediction and protein function prediction.

Chapter 1 - On the basis of achievements of deep neural network technologies, its use in the area of recommendation framework has gained a lot of attention. This chapter provides usage of various deep learning techniques used in recommendation models. The ability of various frameworks based on deep neural networks analyze the user-item relationship efficiently is the reason to use it instead of the traditional recommendation models. This chapter consists of the introduction of basic terms and concepts in both recommendation system and deep learning. Then there is a description of research that specifies various applications of frameworks based on deep neural networks in the field of recommendation system.

Chapter 2 - Text recognition has risen in popularity in the area of computer vision and natural language processing due to its use in different fields. For character recognition in a handwriting recognition system, several methods have been suggested. There are enough studies that define the techniques for translating text information from a piece of paper to an electronic format. Text recognition systems may play a key role in creating a paper-free environment in the future by digitizing and handling existing paper records. This chapter provides a thorough analysis of the field of Text Recognition.

Chapter 3 - Diabetic Retinopathy (DR) is one of the common issues of diabetic Mellitus that affects the eyesight of humans by causing lesions in their retinas. DR is mainly caused by the damage of blood vessels in the tissue of the retina, and it is one of the leading causes of visual impairment globally. It can even cause blindness if not detected in its early stages. To reduce the risk of eyesight loss, early detection and treatment are pretty necessary. The manual process by ophthalmologists in detection DR requires much effort and time and is costly also. Many computer-based techniques reduce the manual effort, and deep learning is used more commonly in medical imaging. This chapter will discuss deep learning and how it is helpful in the early detection and classification of DR by reviewing some latest state-of-art methods. There are various datasets of colour fundus images available publically, and the authors have reviewed those databases in this chapter.

Chapter 4 - Breast cancer is a type of cancer that develops in the cells of the breast and is a fairly prevalent disease in women. Breast cancer, like lung cancer, is a life-threatening condition for women. A promising and significant tool is automated computer technologies, particularly machine learning, to facilitate and improve medical analysis and diagnosis. Due to the great dimensionality and complexity of this data, cancer diagnosis using gene expression data remains a challenge. There is still ambiguity in the clinical diagnosis of cancer and the identification of tumor-specific markers, despite decades of study. In this study, the authors discuss various feature extraction techniques on different kinds of datasets. The authors also discuss various deep learning approaches for cancer detection and the identification of genes important for breast cancer diagnosis.

Chapter 5 - Deep learning helps simulation techniques with various computing layers to gain several stages of abstraction for data representations. These techniques have vastly enhanced the position in voice detection, visual target recognition, particle identification as well as a variety of other fields including drug discovery as well as genomics. Deep learning uses the backpropagation method to show how a computer can adjust the input variables that are employed to measure the value in every layer from the description in the subsequent layer revealing detailed structure in huge volumes of data. Deep learning has accomplished substantial progress as well as demonstrated outstanding effectiveness in a variety of applications like Adaptive testing, cancer detection, natural language processing, face recognition, speech recognition, and much more. Epileptic seizures are chronic neurological disorders with serious public health consequences. Epileptic seizures are a type of neurological disorder that can affect children between the ages of 10 to 20, as well as adults between the ages of 65 to 70. Pre-ictal, Ictal, Post-ictal, and Interictal epileptic seizures are the four phases that can be analyzed. The basic goal of this article is to offer a sequential overview of the primary uses of deep learning in a range of fields, including a study of the methodologies and architectures employed as well as the impact of each implementation in the actual world. This chapter shows that ANN had 98.26% accuracy, CNN had 97.52% accuracy, and LSTM had 95.78% accuracy. In terms of F1-Score, the authors found that ANN has a score of 95.57%, CNN has a score of 93.77%, and LSTM has a score of 89.87%. In addition, when compared to previous techniques, the convolutional neural networks model performed better.

Chapter 6 - Deep learning (DL), also known as hierarchical learning or deep structured learning, is a sub-domain of machine learning. The term deep in the expression deep learning refers to the number of layers in a neural network. DL has shown a significant breakthrough in many areas with outstanding performance. After seeing the performance, DL has been used in high dimensional data and many complex applications, giving state of-the-art results compared to conventional methods. In this paper, some applications are discussed where deep learning is widely used, namely healthcare, self-driving or autonomous cars, NLP, speech recognition, image recognition, cybersecurity, Automatic coloring, etc. Many areas are adopting deep learning. Also, the authors have discussed some of the DL architectures commonly used in the applications mentioned above. Deep learning is becoming more popular because of the large data available for training the networks, deep architectures, activation functions, and more computational power. And deep architectures have many hidden layers that are more powerful and useful for robust feature learning to increase performance.

Chapter 7 - Examination of public opinion (also called opinion mining or sentiment analysis (SA)) is a hot topic in the natural language processing (NLP) field. The SA aims to define, extract, and organize sentiments from user-generated texts in places like social media, blogs, news headlines, and reviews on product. Most of the researchers in the literature have used machine learning methods to address SA tasks from various perspectives over the last two decades. Since an NLP researchers' output is heavily influenced by data representation choices, many studies focus on developing powerful feature extractors using domain knowledge and careful engineering. Deep learning techniques have recently emerged as effective computational frameworks capable of automatically discovering complex semantic representations of texts from data without the need for function engineering. Many SA activities, such as sentiment recognition, opinion extraction, fine-grained SA, and others, have benefited from these methods. In this work, the authors provide an overview of effective deep learning approaches to SA activities at various scales.

Chapter 8 - Protein-Protein Interactions are essential in all organisms cellular functioning and biological processes. The majority of genes and proteins realize the resulting phenotype function as a set of interactions. protein-protein interactions are an important molecular process in cells and play a key role in their biochemical function. Where as high-throughput experimental techniques have advanced, allowing researchers to detect large quantities of protein-protein interaction, they are not without drawbacks, deep learning techniques have been widely used to predict protein-protein interaction, thus allowing experimental researchers to study protein-protein interaction networks. This paper provides a brief study of methods used in Protein-Protein Interactions, their limitations, and their applications of Protein-Protein Interactions.

Chapter 9 - Today's majority of IT industries/companies demand automation. Software is the most important component for automation. Many companies or industries work for software development for specific or universal purposes. During the software development process, Software Defect Prediction (SoDP) is also an important term to use. In the active research areas of software engineering, SoDP plays a significant role. SoDP is finding the errors or faults or bugs or defects in software before deployment the software. For automation, the SoDP process is an important part of it. This chapter mainly focuses on various machine learning techniques, which are used for the prediction of defects in software. Most of the researchers are used supervised learning techniques to defect prediction in software. Similarly, many researchers also based on unsupervised learning techniques for the defect prediction in software. At present, the majority of researchers are focused on reinforcement learning techniques due to automation.

LIST OF REVIEWERS

Dr. Akshay Deepak	NIT Patna
Dr. Shailendra Tiwari	TIET Patiala
Dr. Sudhakar Tripathi	REC Ambedkar nagar
Dr. Anand Shankar Tiwari	NIT Patna
Dr. Durgesh Singh	IIITDM Jabalpur
Dr. Rajesh Kumar	Allahabad University, Prayagraj
Dr. Vaibhav Soni	MANIT Bhopal
Dr. J.P.Singh	NIT Patna
Dr. N.P.Singh	NIT Hamirpur
Dr. Vibhav P. Singh	MNNIT Allahabad

In: Deep Learning and Its Applications ISBN: 978-1-68507-185-1
Editor: Arvind Kumar Tiwari

Chapter 1

APPLICATION OF DEEP LEARNING IN RECOMMENDATION SYSTEM

Ishwari Singh Rajput[1], Anand Shankar Tiwari[1] and Arvind Kumar Tiwari[2]
[1]CSE Department, NIT Patna, Patna, Bihar, India
[2]CSE Department, KNIT Sultanpur, Sultanpur, Uttar Pradesh, India

ABSTRACT

On the basis of achievements of deep neural network technologies, its use in the area of recommendation framework has gained a lot of attention. This chapter provides usage of various deep learning techniques used in recommendation models. The ability of various frameworks based on deep neural networks analyze the user-item relationship efficiently is the reason to use it instead of the traditional recommendation models. This chapter consists of the introduction of basic terms and concepts in both recommendation system and deep learning. Then there is a description of research that specifies various applications of frameworks based on deep neural networks in the field of recommendation system.

Keywords: deep neural network, recommendation system, autoencoder

INTRODUCTION

The advancement in the field of information technology in recent times has made the access of huge amounts of data very easy. There are a vast variety of products and services whose description is easily available along with their comments and reviews. Due to this overload of information [1], it becomes very difficult for the user to choose an appropriate product according to his requirements.

To deal with the above problems, recommendation systems are used. Recommendation Systems provide the users with personalized recommendations. There are many areas where recommendation system is being used, such as music, books, movies, shopping etc. Most of the online vendors have a recommendation engine already equipped. Recommendations are done on the basis of user's previous items choice or the items preferred by similar user or on the basis of the description of item. Here the item refers to the product or services which is to be recommended. Recommendation system is classified into three parts based on their approach: [2] Content based, collaborative filtering and hybrid.

Recently, the use of artificial neural networks has become very popular in the problems which require complex computations and huge amount of input data. Deep learning is a part of ANN architectural models of deep neural network are efficiently built and trained. Deep neural networks have its applications in various fields such as speech recognition, image processing, object recognition, image processing, NLP tasks etc. Due to various advantages of deep learning, researchers have been encouraged to use its associated techniques in the field of recommendation system also.

Deep learning is being used successfully in recommendation system as well as many other fields in computer science and has shown significant improvements in the existing models. In 2007, a collaborative filtering method for movie recommendation system was given by [3] which utilized the hierarchical model of deep learning. In 2015, [4] with the use of auto encoders, predicted the values which were missing in the user-item matrix.

The sparsity issue in recommendation system in collaborative filtering was addressed by [5].Various surveys have been done in the area of deep neural network based recommendation system. The state-of-the-art survey for deep recommendation system has been done by [6]. In 2017, a comprehensive review on deep learning based recommendation system has been done by [7]. The paper proposed the classifications on the basis of their structure that is neural network models and integration models.

This article is organized in various sections as follows. In the first section, introduction of recommendation and deep neural network based techniques are introduced. Section two consists of background and related terminologies. Section 3 describes the usage of various approaches of deep neural networks in the area of recommendation system and its classification.

Background and Terminologies

A recommendation system is used for information filtering and outputs a list of specific products in a personalized manner. For examining how the two fields that are recommendation system and deep learning are integrated together, one should know the basics of both these fields. This section of the paper includes a brief description about the fundamental classifications and challenges of both the fields. First, there is an introduction of types of recommendation system and then the details about deep learning methods.

Recommendation System

Various recommendation system approaches are evolved over time and applied in various applications.User finds to filter the huge amount of data available as important task, to find a useful, tailored and relevant content. The recommendations that are predicted by the recommendation System helps the users in taking a decision.

In a traditional recommendation system, the recommendations can be made in two different manners, i.e., predicting a particular item or preparing a ranking list of items for a particular user [8]. The recommendation model is categorized into Content based model [9], Collaborative filtering based model and hybrid recommendation framework [10].

1) Content based recommendation framework: In this model, the items which are same in content is searched. The profile of user is established on the basis of items on which the user is interested in. According to the profile generated, the recommendation system searches the database for the appropriate items using the descriptive attributes of the item. If we use this recommendation system [11] for an item which is newly added, then content based recommendation system works very efficiently. The problem with new inserted item is that it may not have any rating, but still the algorithm works since it uses descriptive information for recommendations. The limitation of this method is that it cannot recommend diverse range of products since the algorithm does not take the information from similar users.
2) Collaborative filtering recommendation systems: This recommendation system assumes that users', who have previously preferred same items, would have same choice in future also. In this system, the recommendations are done on the basis of similar users' pattern rather than descriptive features of items. A correlation among the users is determined, depending upon the choice of similar users, the items are recommended.

There are two methods which are followed by Collaborative filtering algorithm: memory based algorithm and model-based methods.

In memory based algorithm the complete user-item matrix is taken into consideration for identifying similarity. After finding out the nearest neighbor, on the basis of neighbors past rating, the recommendations are provided. In model based algorithms, an offline model is built with the use of machine learning algorithms and data mining methods. These models include clustering models, Decision models, Bayesian model and singular value decomposition model.

3) Hybrid Recommendation Framework: It is an integration of content-based model and collaborative filtering model; it incorporates the benefits of both the methods and tries to eliminate the limitations of the above models [12].

   There are many hybrid systems proposed, some of them are as following:

   Cascade: The output of one approach is later on used by other approach

   Switching: Recommendation output is given on the basis of current situation and either of the one approach is used.

   Weighted: the combination of the scores of various approaches is used for recommendation.

Deep Learning Techniques

Deep learning is based on learning many layers of representations with the help of artificial neural networks, and is a part of machine learning. Deep learning has its applications in variety of fields like Computer vision, recognition of speech, natural language synthesis etc. The important factors which increase the importance of deep neural network as the state-of-the-art machine learning methods are as following:

Big data: As the amount of data increases, better representations are learnt by the deep learning model.

Computational power: The complex computations of the deep learning model is done using the GPU

In this section, there is a description of various deep learning models which are used in recommendation systems.

Autoencoder

Autoencoder is an unsupervised learning technique in which neural networks are applied for the task of *representation learning*. Specifically, a neural network architecture is designed which imposes a bottleneck in the network which forces a *compressed* knowledge representation of the original input as shown in figure 1. In autoencoder some representations from the encoded input are found by the training, so that the input can be restored back from these representations. Autoencoder have three layers arranged as the input, hidden and output layer. There are equivalent numbers of neurons in the input and output layer. The compressed representations are obtained from the hidden layer, and with the help of these representations, the first layer inputs are reconstructed at the output layer of autoencoder [13].

In the learning process, two mappings are used, with the help of encoder and decoder. An encoder is fully connected deep neural network which transforms the input into a latent space representation. Decoder is also having a similar structure as encoder and it is responsible for reconstructing the input back to the original form from the hidden layer outputs [14]. There are many variants of autoencoders such as sparse autoencoder, stacked autoencoder, denoising autoencoder and variational autoencoder etc. Data Compression, Image Denoising, Dimensionality Reduction, Feature Extraction, Image Generation and Image colorization are few important applications of autoencoders.

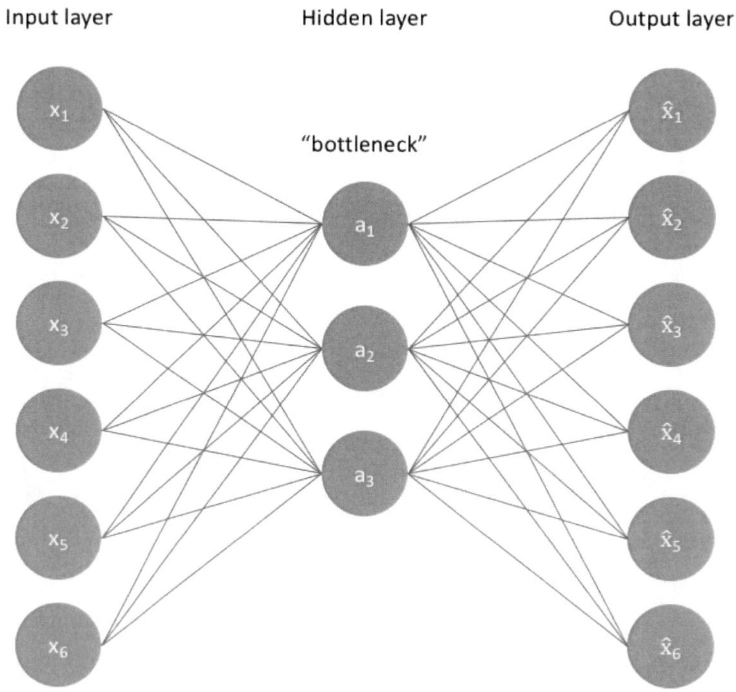

Figure 1. Autoencoder.

Recurrent Neural Network

Sequence prediction issues have been around for quite a while. They are considered as probably the most difficult issue to settle in the information science industry. These incorporate a wide scope of issues, from foreseeing deals to discovering designs in financial exchanges. In the ordinary feed-forward neural organizations, all experiments are viewed as free. That is when fitting the model for a specific day; there is no thought at the stock costs on the earlier days. This reliance on time is accomplished by means of Recurrent Neural Networks. An ordinary RNN structure glances as demonstrated in figure 2.

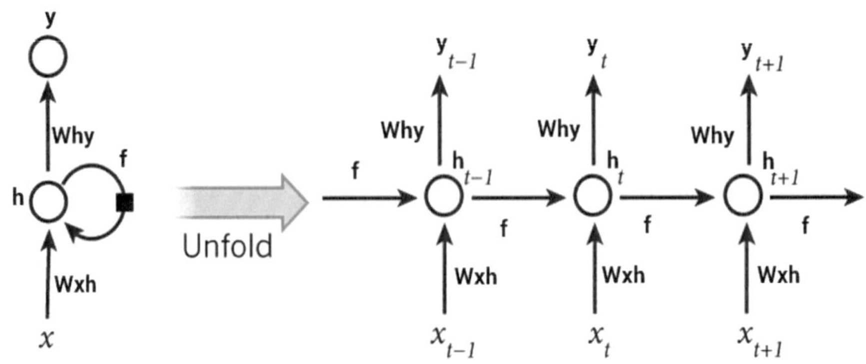

Figure 2. Recurrent Neural Network.

In RNN, the sequence of values, i.e., $x^{(0)}$, $x^{(1)}$,..., $x^{(t)}$ is processed. On each element of the sequence, the same task is performed and the output is based on previous computations. RNN [15] uses an internal memory to hold the values of previous computations, so that it may be used later. Recurrent Neural Networks work moderately okay when we are dealing with short-term dependencies. But RNN fails to deal with large term dependencies. The reason behind this is the problem of Vanishing Gradient and exploding gradient. To remove this problem, Long Short Term Memory (LSTM) and gated recurrent unit is used. It is used when the processing is to be done for predicting events which has comparatively longer interval and delays. LSTM has a processor to distinguish the useful information from the information which are not useful. This processor is known as cell. There are three gates in LSTM namely input gate, output gate and forget gate. The structure of LSTM is shown below in figure 3.

A forget gate is answerable for eliminating data from the cell state. The data that is not, at this point needed for the LSTM to get things or the data that is of less significance is taken out by means of augmentation of a channel. This is needed for advancing the exhibition of the LSTM organization. The info entryway is answerable for the expansion of data to the cell state.

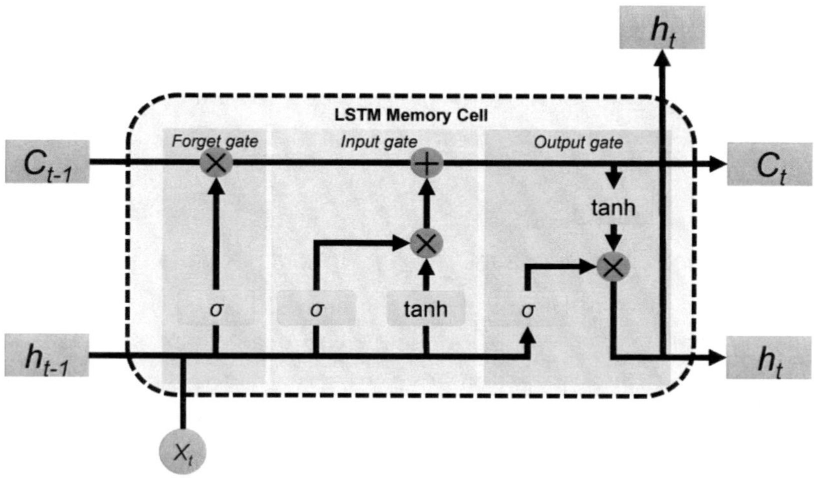

Figure 3. Architecture of LSTM Cell.

The work of choosing helpful data from the current cell state and showing it out as a yield is done by means of the yield gate.RNN are valuable while managing worldly elements of communications and when the client's conduct has a specific arrangement of pattern.

Convolution Neural Network

A CNN or convolutional neural network [16] is feed-forward neural organization that is by and large used to dissect visual pictures by preparing information with matrix like geography. It's otherwise called a ConvNet. A convolutional neural organization is utilized to recognize and characterize objects in a picture. A convolution neural organization has numerous secret layers that help in extricating data from a picture. The four significant layers in CNN are:

1. Convolution layer
2. ReLU layer
3. Pooling layer
4. Fully connected layer

This is the initial phase during the time spent separating significant highlights from a picture. A convolution layer has a few channels that play out the convolution activity. ReLU represents the corrected direct unit. When the component maps are extricated, the subsequent stage is to move them to a ReLU layer. ReLU plays out a component insightful activity and sets every one of the negative pixels to 0. It acquaints non-linearity with the organization, and the produced yield is an amended element map. Pooling is a down-inspecting activity that decreases the dimensionality of the component map. The corrected component map presently goes through a pooling layer to create a pooled highlight map. The following stage in the process is called straightening. Leveling is utilized to change over every one of the resultant 2-Dimensional exhibits from pooled highlight maps into a solitary long consistent direct vector. The smoothed lattice is taken care of as contribution to the completely associated layer to arrange the picture. There are many applications in which CNNs are applied such as object recognition, self-driving cars, audio processing etc. while transforming the input to output. The structure of the typical CNN model is shown in figure 4.

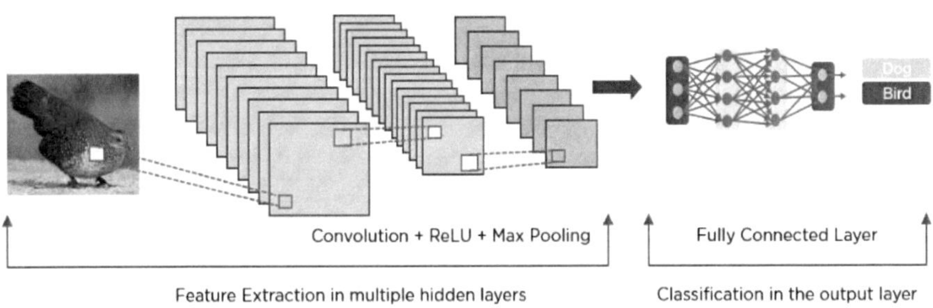

Figure 4. Convolutional Neural Network model.

Restricted Boltzmann Machine

RBM is developed by Geoffrey Hinton, and are stochastic neural organizations that can gain from a likelihood conveyance over a bunch of sources of info. It comprises of two layers of neural organization in particular an obvious and a secret layer. Each noticeable unit is associated with all secret units. RBMs have a predisposition unit that is associated with every one of the noticeable units and the secret units, and they have no yield hubs. The construction of the RBM model is appeared in figure 5.

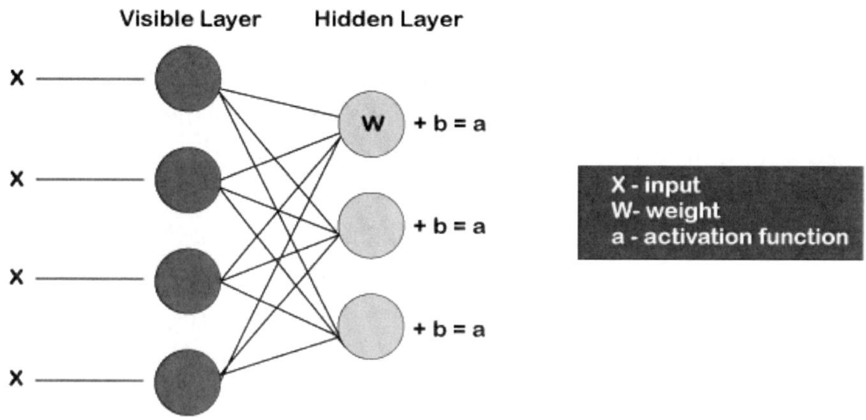

Figure 5. Restricted Boltzmann Machine model.

The complex computations and learning in RBM [17] is based on characteristic articulation of information. The word "restricted' is used for intra-layer communication as it is not present in both hidden layer and visible layer. Due to this restriction, the learning efficiency increases. There is a full connection between the nodes of different layers which are stacked together and there is no connection between the nodes of same layer. Applications of RBM includes dimensionality decrease, characterization, relapse, community separating, highlight learning, and theme displaying. Since RBM utilizes a basic forward encoding activity, so it is quick when contrasted with different models, for example, autoencoder.

APPLICATION OF DEEP LEARNING IN RECOMMENDATION SYSTEM

DL techniques are widely used in various real world applications such as sentiment analysis, speech recognition image classification, text classification etc. Various researchers also include deep neural network based techniques in the field of recommendation system to improve its performance as compared to traditional recommendations systems. Traditional recommendation methods use matrix factorization methods which have some limitations such as:

1. The trouble of utilizing side highlights that may influence the suggestion like the U/PG rating of a film or the nation of the client. We can just utilize the thing ID and the client ID on account of network factorization. It likewise keeps us from questioning a thing or client not present in the preparation set.
2. The matrix factorization additionally had the virus start issue because of the way that it had no component vector or inserting for the new things.
3. Matrix factorization regularly will in general prescribe mainstream things to everybody which doesn't generally mirror the particular client interests for the most part when Dot items are utilized.
4. Matrix factorization deals with the straightforward internal result of the User and thing highlight embeddings, it is regularly insufficient to catch and address the mind boggling relations in the client and things.

Deep Neural networks are planned and used to address these weaknesses of the matrix factorization strategies. This section describes the various categories recommendation systems which are based on deep learning.

The categorization is based on the types of recommendation systems used which is as follows:

1. Collaborative filtering recommendation system based on deep neural network.
2. Content-based recommendation system based on deep neural network.
3. Hybrid recommendation system based on deep neural network.
4. Social network-based recommendation system based on deep neural network.
5. Context aware recommendation system based on deep neural network.

Hybrid model and neural network model are the two categories of deep neural network-based recommendation system. Integration model is further divided into two categories on the basis of whether it combines any traditional recommendation system model with deep neural network technique or depends solely on deep learning method.

Neural network model is also divided into two categories on the basis of deep neural network based technique used: models which uses single deep neural network based technique and deep neural network based composite model. In deep neural network based composite model, different deep neural network techniques are used to build a hybrid system having more capability.

1. Collaborative Filtering Recommendation Systems Based on Deep Neural Networks

Collaborative filtering (CF) is one of the usually implemented techniques in recommendation systems in order to tackle various real-life issues. The state of the art CF-based methods uses the rating matrix for recommending the items. But this approach faces the problem of data sparseness and cold start problem. The sparsity of the user-item matrix, the learned features is not effective which reduces the performance of recommendation system. Various researchers propose deep neural network based collaborative filtering techniques to enhance its effectiveness in recommendation.

1.1. Collaborative Filtering Method Based on Generative Adversarial Network

Generative Adversarial Network is a neural network which is generative and having discriminator and generator functions. These both functions are simultaneously trained in competition with one another in architecture of minimax game. The first model to implement GAN in the field of Information Retrieval is (IRGAN) [18] which stands for Information retrieval generative adversarial network. The state of the art GAN model has two modules a discriminator and a generator. The generative retrieval module predicts appropriate documents with given query, whereas discriminative retrieval module predicts relevancy given with a pair of query and document.

The IRGAN model combines above two Information Retrieval models in order to play a minimax game with them: the generative retrieval model produces (or selects) relevant documents that are relevant documents like ground truth, while the discriminating retrieval model separates the relevant documents from those generated by the generative retrieval model [32]. concentration is on the semantic-rich client thing communications in a recommender framework and propose a novel generative adversarial network (GAN) named Convolutional Generative Collaborative Filtering (Conv-GCF). They build up a powerful irritation system (ill-disposed commotion layer) for convolutional neural organizations (CNN), in light of which a generator is planned with lingering squares to combine client thing collaborations.

1.2. Recurrent Neural Network Based Collaborative Filtering Method

In order to deal with the information in sequential form, recurrent neural network (RNN) proves to be a very effective network. Concepts of loops are used in place of feedforward network to remember sequences.

Variants of RNN viz. Long Short-Term Memory (LSTM) and Gated Recurrent Unit (GRU) network are used to deal with the problems of long term dependencies and vanishing gradient problem. In collaborative filtering method based on RNN, the impact of user historical pattern is modelled on the current behavior of user, recommendation is performed and user's behavior is predicted [19]. Figure 7 shows the framework of collaborative filtering method based on RNN [19]. Let the input set is $\{I_1, I_2 . . . I_t\}$, and output is $O_t = \sigma\ (f\ (W \cdot h_{t-1} + V \cdot I_t) \cdot V)$, σ represents a softmax function, f represents the activation function, which specifies the selection probability of any item at time t. h_t represents the hidden state vector.

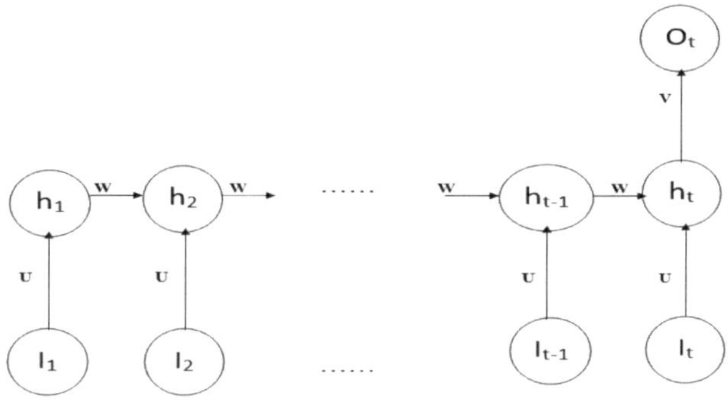

Figure 6. Collaborative filtering model based on RNN.

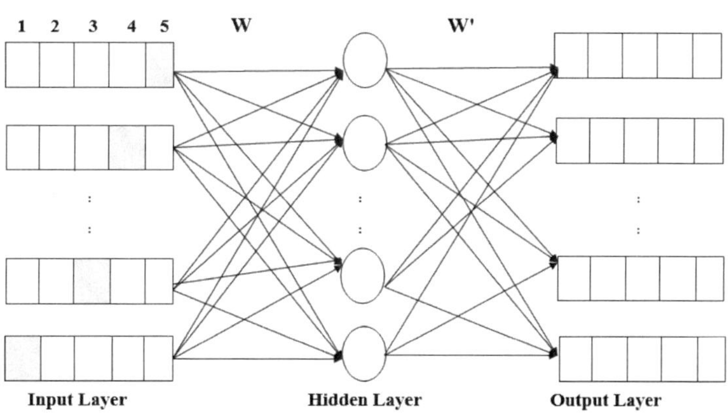

Figure 7. Collaborative filtering method based on Autoencoder.

1.3. Collaborative Filtering Method Bssed on Autoencoders

The first ever developed autoencoder-based collaborative recommendation model is Autoencoder based Collaborative filtering [20]. It decomposes the vectors by integer ratings. The model proposed by [20] takes client or thing based evaluations as contributions to rating matrix R. The output is produced by the process of encoding and decoding by optimizing the parameters of model and reducing the reconstruction error. Consider an example, if the range of integers [1-5] represents the rating score, then each r_{ui} can be divided into five vectors.

Above figure represents the 1 to 5 rating scale in which blue boxes represents the user rated item. The cost function which is to be reduced is taken as Mean Square Error. The rating prediction in this approach is found by making the summary of each of the five vectors, which are scaled by rating K. Pretraining of parameters and local optimum avoidance is performed by RBM. Stacking multiple autoencoder collectively shows the slight improvement in accuracy. This method based on autoencoder suffers from the problem of dealing with non-integer ratings and sparseness of input data due to decomposition of partial observed vectors.

Collaborative Denoising Auto-Encoder [21] is primarily used for prediction rankings. User feedback is taken as input to the CDAE. If the user enjoys a movie, the input value is 1 otherwise it is 0. It shows the vector preference to display the user's interest in some item. Gaussian noise corrupts the CDAE input.

1.4. Collaborative Filtering Method Based on Restricted Boltzmann Machine

Restricted Boltzmann Machine (RBM) is a double-layer neural network capable to deal with typical learning based problems. The efficiency of learning is improved by removing the connections between same layers.

This recommendation method is proposed by [22]. Further a conditional RBM model is proposed to consider information in form of feedback. The visible layer of RBM can take only binary values so only one hot vector can be used to represent the rating score. The architecture of RBM based model is represented as shown in Figure 8.

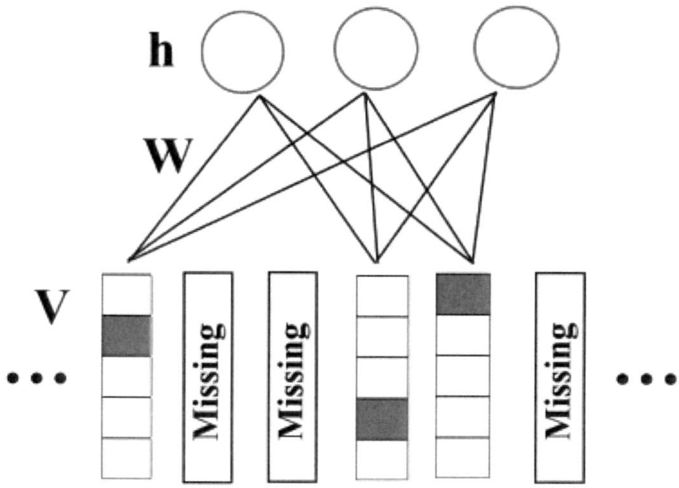

Figure 8. Collaborative filtering model based on RBM.

There are equal hidden layers in each RBM and have softmax units which are visible for movie ratings given by user. There is one training case in each RBM unit with combined weights and biases. Hidden units hold the binary states which are different for separate users. Let there are *N* movies rated by user and there are *n* visible units. Suppose *M* represents $X \times N$ size matrix where $r_{ui}^{x}= 1$ if movie *i* is rated as x by user *u,* otherwise $r_{ui}^{x}= 0$. So,

$$p(r_{ui}^{x} = 1|h) = \frac{\exp(rb_i^x + \sum_{j=1}^{T} h_j j w_{ij}^x}{\sum_{l=1}^{X} \exp(b_i^l + \sum_{j=1}^{T} h_j w_{ij}^l)} \quad (1)$$

$$p(h_j = 1|\mathrm{M}) = \sigma\,(\mathrm{b_j} + \sum_{i=1}^{N} \sum_{k=1}^{T} r_{ui}^{x}\, \mathrm{w_{ij}^{x}}) \quad (2)$$

where σ (x) represents a logistic function, r_{ui}^{x}: interaction, b_i^x :bias of rating *x* for movie *i* and b_j: hidden unit bias.

2. Content-Based Recommendation Systems Based on Deep Neural Networks

Content-based recommendation systems recommend on the basis of descriptive attributes of items and users' profiles such as texts, pictures and videos [23]. The use of deep neural network in content-based recommendation systems is to capture the non-linear relationships between user and item [24]. It also captures the intra-relationship between data, from large available data sources.

3. Hybrid Recommendation System Based on Deep Neural Networks

The state-of-the-art Collaborative Filtering-based approach uses the ranking matrix to suggest products. But this method suffers from a data sparsity and a cold start crisis. Due to the scarce existence of the user-item matrix, the features learned are not accurate, which decreases the efficiency of the recommendation framework. Hybrid recommendation model based on deep neural networks incorporates a content-based recommendation model with collective recommendation-based filtering models, which combines the mechanism of feature learning and recommendation into a single model.

[25] suggested a layered, self-encoder-based hybrid paradigm that learns the latent space factors of users and items and simultaneously performs mutual filtering from the ranking matrix.

An autoencoder is a variation of neural organization, having encoder and decoder as two components. The encoder changes over the contribution to its secret portrayal, while the decoder changes over the secret portrayal back to the re-established input structure. The boundaries comparing to the autoencoder are prepared to limit the mistake because of the recreation, which is estimated by the loss work or misfortune work. Denoising autoencoder (DAE) attempts to recreate the contribution from a ruined adaptation for improved portrayal from the info. More variations of autoencoder have been created for better results. The crossover suggestion model dependent on the stacked denoising autoencoder utilizes both the rating network and the side data and coordinates both the SDAE [26] and the grid factorisation. Lattice factorization is a broadly utilized model with improved exactness, and SDAE is an incredible model for separating significant level highlights from inputs. The combination of the over two model will turn into an amazing model for additional advantages.

4. Social Network-Based Recommendation System Using Deep Neural Networks

Conventional recommendation models never consider social connections among the users. But we always take verbal recommendations from our friends in reality.These verbal recommendations are termed as social recommendation which occurs daily [27]. Hence, for improved recommendation systems and for more personalized recommendations, social network must be employed among users. Every user will interact with various types of social relationships.

The quality of recommendation system is very crucial task which can be achieved by implementing the effect of social relationship among the users. Items with location attributes and sequential pattern of user behaviour in spatial and temporal frame are used to form spatiotemporal pattern which is used to improve recommendation accuracy. Recently, a very few recommendation techniques have been proposed which is based on the users' trust relations improve conventional recommendation systems. These trust based recommendation models proves to be an effective move in the field of recommendation system models.

In current scenario an integration of deep learning and social network based recommendation system provides a platform for various research solutions. The limitations which are inherent to the social recommendation must be addressed in the future research.

5. Context-Aware Recommendation Systems Based on Deep Neural Networks

A context-aware recommendation system, integrates context based information into a recommendation model. This integration is effectively performed by the deep learning techniques in different conditions of recommending items [28]. Deep neural network based methods are used to extract the latent space presentation from the context based information. Deep learning based model can be integrated into diverse data to reduce data sparsity problem [29].

Sequential nature of data plays a significant part in implementing user behaviours. Recently, recurrent neural networks (RNNs) are commonly used in a variety of sequential simulation activities. However, for real-world implementations, these approaches have trouble modeling contextual knowledge, which has been shown to be very essential for behavioural modelling.

Currently this method based on deep neural networks focused towards situation information. A novel approach is proposed called context-aware recurrent neural networks. It uses two types of matrices: input and transition matrices. They both are specific to the context and adaptive in nature. Input matrices are used to extract various situations such as time, place, weather condition where actually the user behaves.

6. Applications

As deep learning plays a significant role in most of the fields as it has the capability of dealing with large and complex problems with improved results. Deep learning technology also contributes in the field of recommendation system for improved customer satisfaction. Deep learning technology overcomes the shortcomings of traditional models to get high quality recommendations. [30].

All the above discussed methods of recommendation systems use deep neural networks and hence also achieve the quality in recommending items to the users [31]. Different recommendation models use different deep learning methods to obtain improve results.

The summary of various deep neural network based recommendation systems are given below:

Application of Deep Learning in	Findings
Content-based recommendation systems	Extract non-linear user-item relationship and intricate relationship with data itself.
Collaborative filtering recommendation systems	Takes the user-item interaction matrix as information and utilizations a profound neural netwrok based model to become familiar with the inactive space introduction that relates to user or items. Utilize the misfortune capacity to build a profound neural network based model improvement work. Based on the dormant space introduction, proposals are made.
Hybrid recommendation systems	Integrate the individual or object learning process and the suggestion process into a single architecture.
Social network-based recommendation systems	Focuses on social relationship between users, and extracts the effect of location of user, movement patterns and other various factors.
Context-aware recommendation systems	Deep neural network based methods combines the context information into the recommendation model and obtain the latent space presentation of the context based information. Also reduce data sparsity in model.

REFERENCES

[1] Gantz, J. and Reinsel, D. "The digital universe in 2020: Big data, bigger digital shadows, and biggest growth in the far east", in *Proc. IDC iView, IDC Analyze Future,* 2012, pp. 1 - 16.

[2] Çano and Morisio, M. "Hybrid recommender systems: A systematic literature review", *Intell. Data Anal.,* vol. 21, no. 6, pp. 1487 - 1524, 2017.

[3] Salakhutdinov, R., Mnih, A., Hinton, G. E. (2007). Restricted Boltzmann machines for collaborative filtering. In: *Proceedings of the 24th international conference on machine learning, Corvallis, Oregon, USA,* pp. 791 - 798.

[4] Sedhain, S., Menon, A. K., Sanner, S., Xie, L. (2015). Autorec: autoencoders meet collaborative filtering. In: *Proceedings of the 24th international conference on world wide web, Florence, Italy,* pp. 111 - 112.

[5] Wang, H., Wang, N., Yeung, D. Y. (2015b). Collaborative deep learning for recommender systems. In: *Proceedings of the 21th ACM SIGKDD international conference on knowledge discovery and data mining, Sydney, NSW, Australia*, pp. 1235 - 1244.

[6] Zheng, L. (2016). *A survey and critique of deep learning on recommender systems.* Technical report, University of Illinois.

[7] Zhang, S., Yao, L., Sun, A. (2017a). Deep learning based recommender system: A survey and new perspectives. *arXiv: 1707.07435.*

[8] Park, D. H., Kim, H. K., Choi, I. Y., Kim, J. K. A literature review and classification of recommender systems research. *Expert Syst. Appl.*, 39(11), 10059 – 10072, (2012).

[9] Verbert, K. et al. "Context-aware recommender systems for learning: A survey and future challenges", *IEEE Trans. Learn. Technol.,* vol. 5, no. 4, pp. 318 - 335, Oct./Dec. 2012.

[10] Mooney, R. J. and Roy, L. "Content-based book recommending using learning for text categorization", in *Proc. 5th ACM Conf. Digit. Libraries, San Antonio, TX, USA,* 2000, pp. 195 - 204.

[11] Lops, P., de Gemmis, M. and Semeraro, G. "Content-based recommender systems: State of the art and trends", in *Recommender Systems Handbook.* Boston, MA, USA: Springer, 2011, pp. 73 - 105.

[12] Tran, T., Cohen, R (2000). Hybrid recommender systems for electronic commerce. In: *Proceedings of the AAAI workshop on knowledge-based electronic markets, Austin, TX, USA,* vol. 4.

[13] Deng, S., Huang, L., Xu, G., Wu, X., Wu, Z. (2017). On deep learning for trust-aware recommendations in social networks. *IEEE Trans. Neural. Netw. Learn. Syst.,* 28(5):1164 - 1177.

[14] Strub and Mary, J. "Collaborative filtering with stacked denoising autoencoders and sparse inputs", in *Proc. NIPS Workshop Mach. Learn. eCommerce, Montreal, QC, Canada*, 2015, pp. 150 - 158.

[15] Wu, C., Wang, J., Liu, W. and Liu, J. "Recurrent neural network based recommendation for time heterogeneous feedback", *Knowl.-Based Syst.,* vol. 109, pp. 90 - 103, Oct. 2016.

[16] Gong, Y. and Zhang, Q. "Hashtag recommendation using attention-based convolutional neural network", in *Proc. 25th Int. Joint Conf. Artif. Intell., New York, NY, USA,* 2016, pp. 2782 - 2788.

[17] Georgiev, K., Nakov, P. (2013). A non-iid framework for collaborative filtering with restricted Boltzmann machines. In: *Proceedings of the 30th international conference on machine learning,* pp. III–1148–III– 1156.

[18] Wang, J. et al. "IRGAN: A minimax game for unifying generative and discriminative information retrieval models", in *Proc. 40th Int. ACM SIGIR Conf. Res. Develop. Inf. Retr., Tokyo, Japan,* 2017, pp. 515 - 524.

[19] Wu, C., Wang, J., Liu, W. and Liu, J. "Recurrent neural network based recommendation for time heterogeneous feedback", *Knowl.-Based Syst.,* vol. 109, pp. 90 - 103, Oct. 2016.

[20] Sedhain, S., Menon, A. K., Xie, L. and Sanner, S. "AutoRec: Autoencoders meet collaborative filtering", in *Proc. 24th Int. Conf. World Wide Web, Florence, Italy,* 2015, pp. 111 - 112.

[21] Strub, F. and Mary, J. "Collaborative filtering with stacked denoising autoencoders and sparse inputs", in *Proc. NIPS Workshop Mach. Learn. eCommerce, Montreal, QC, Canada,* 2015, pp. 150 - 158.

[22] Mnih, A., Hinton, G. and Salakhutdinov, R. "Restricted Boltzmann machines for collaborative filtering", in *Proc. 24th Int. Conf. Mach. Learn., Corvalis, OR, USA,* 2007, pp. 791 - 798.

[23] Van Meteren, R., Van Someren, M. (2000). Using content-based filtering for recommendation. In: *Proceedings of the workshop on machine learning in the new information age, Barcelona, Spain,* pp 47 - 56.

[24] Huang, P.-S., He, X., Deng, L., Acero, A., Heck, L. and Gao, J. "Learning deep structured semantic models for Web search using clickthrough data", in *Proc. 22nd ACM Int. Conf. Conf. Inf. Knowl. Manage, San Francisco, CA, USA,* 2013, pp. 2333 - 2338.

[25] Wang, C. and Blei, D. M. "Collaborative topic modeling for recommending scientific articles", in *Proc. 17th ACM SIGKDD Int. Conf. Knowl. Discovery Data Mining, San Diego, CA, USA,* 2011, pp. 448 - 456.

[26] Dong, X., Yu, L., Sun, Y., Yuan, L., Zhang, F. and Wu, Z. "A hybrid collaborative filtering model with deep structure for recommender systems", in *Proc. 31st AAAI Conf. Artif. Intell., San Francisco, CA, USA,* 2017, pp. 1309 - 1315.

[27] Ma, Zhou, D., Lyu, M. R., King, I. and Liu, C. "Recommender systems with social regularization", in *Proc. 4th ACM Int. Conf. Web Search Data Mining, Hong Kong,* 2011, pp. 287 - 296.

[28] Liu, Q., Wu, S., Li, Z., Wang, L. and Wang, D. "Context-aware sequential recommendation", in *Proc. IEEE 16th Int. Conf. Data Mining, Barcelona, Spain,* 2016, pp. 1053 - 1058.

[29] Unger, M., Bar, A., Rokach, L. and Shapira, B. "Towards latent context-aware recommendation systems", *Knowledge-Based Syst.,* vol. 104, pp. 165 - 178, Jul. 2016.

[30] Xu, Z., Chen, C., Miao, Y., Meng, X. and Lukasiewicz, T. "Tag-aware personalized recommendation using a deep-semantic similarity model with negative sampling", in *Proc. 25th ACM Int. Conf. Inf. Knowl. Manage., Indianapolis, IN, USA,* 2016, pp. 1921 - 1924.

[31] Okura, S., Tagami, Y., Tajima, A. and Ono, S. "Embedding-based news recommendation for millions of users", in *Proc. 23rd ACM SIGKDD Int. Conf. Knowl. Discovery Data Mining, Halifax, NS, Canada,* 2017, pp. 1933 - 1942.

[32] Yuan, F., Yao, L. and Benatallah, B. 2020, October. Exploring Missing Interactions: A Convolutional Generative Adversarial Network for Collaborative Filtering. In *Proceedings of the 29th ACM International Conference on Information and Knowledge Management,* (pp. 1773 - 1782).

In: Deep Learning and Its Applications
Editor: Arvind Kumar Tiwari
ISBN: 978-1-68507-185-1

Chapter 2

Deep Learning Based Approaches for Text Recognition

Shalini Agrahari * ***and Arvind Kumar Tiwari***
Department of CSE, Kamla Nehru Institute of Technology,
Sultanpur, India

Abstract

Text recognition has risen in popularity in the area of computer vision and natural language processing due to its use in different fields. For character recognition in a handwriting recognition system, several methods have been suggested. There are enough studies that define the techniques for translating text information from a piece of paper to an electronic format. Text recognition systems may play a key role in creating a paper-free environment in the future by digitizing and handling existing paper records. This chapter provides a thorough analysis of the field of Text Recognition.

Keywords: deep learning, text recognition, CNN, RNN, LSTM

---

* Corresponding Author's Email: shalini.2562@knit.ac.in.

INTRODUCTION

We are all familiar with the convenience of having an editable text document that can be easily read by a computer and the information can be used for a variety of uses. People always wanted to use the text that is present in various forms all around them, such as handwritten documents, receipts, images, signboards, hoardings, street signs, nameplates, number plates of automobiles, as subtitles in videos, as captions for photos, and in a variety of other ways. However, we are unable to make use of this information because our computer is unable to recognize these texts purely based on their raw images. Hence, researchers around the world have been trying hard to make computers worthy of directly recognizing text by acquiring images to use the several information sources that could be used in a variety of ways by our computers. In most cases, we have no choice but to typewrite handwritten information, which is very time-consuming. So, here we have a text recognition system that overcomes these problems. We can see the importance of a 'Text Recognition System' just by having to look at these scenarios, which have a wide range of applications in security, robotics, official documentation, content filtering, and many more.

Due to digitalization, there is a huge demand for storing data into the computer by converting documents into digital format. It is difficult to recognize text in various sources like text documents, images, and videos, etc. due to some noise. The text recognition system is a technique by which recognizer recognizes the characters or texts or various symbols. The text recognition system consists of a procedure of transforming input images into machine-understandable format [1, 2].

The use of text recognition has a lot of benefits. For example, we find a lot of historical papers in offices and other places that can be easily replaced with editable text and archived instead of taking up too much space with their hard copies. Online and offline text recognition are the two main types of recognition whether online recognition system includes tablet and digital pen, while offline recognition includes printed or handwritten documents [3].

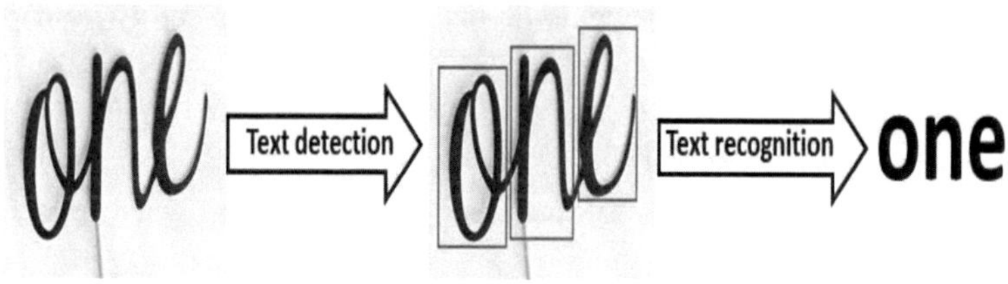

Figure 1. An example of text recognition.

To recognize text in images, the first step is to detect and identify a bounding box around each character of text areas in the image and the second step is to figure out what the characters are. An example of text recognition is given below (See Figure 1).

A text recognition procedure is carried out by some steps such as preprocessing, segmentation, feature extraction, classification, and post-processing [4, 5].

Preprocessing

This initial phase is required to enhance the quality of the input image by doing some operations like noise elimination and normalization, etc. and it is also good for improving recognition rate.

Segmentation

It is a crucial phase in recognition. The process of segmentation is to segment the input image into single characters. It gives separation among the individual characters of the input image.

Feature Extraction

This phase extracts important information from input images by applying different feature extraction techniques such as histograms, etc.

Feature extraction aims to represent data effectively by which recognition rate increases.

Classification

This phase is the decision-making phase which compares the extracted input feature to the stored pattern, and assigns them into the correct character class. ANN or SVM classification techniques are used as trained classifiers.

Post-Processing

This last phase improves the recognition rate by filtering and correcting the output obtained by the classification phase. Some post-processing operations are formatting, spell-checking, and correction of data.

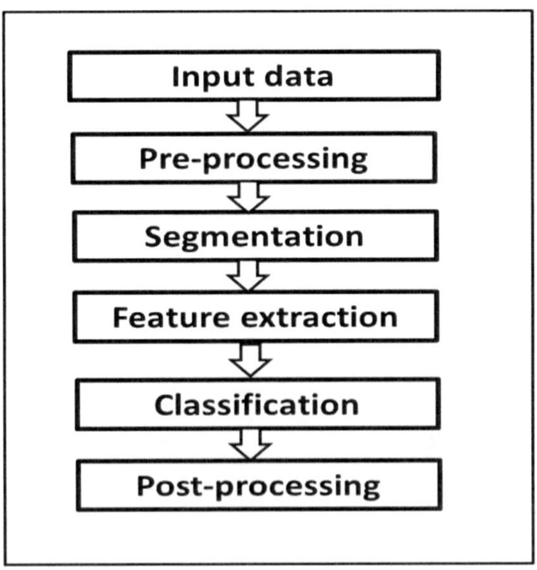

Figure 2. A general diagram of text recognition steps.

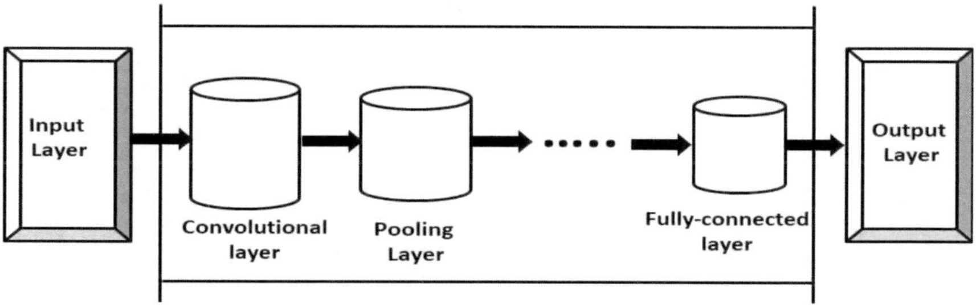

Figure 3. A general CNN diagram.

Many machine learning algorithms, deep learning algorithms like CNN, RNN, and datasets are used in text recognition and detection. The idea of developing a predictive model based on experience is involved in machine learning. The deep neural network can model convoluted and non-linear connections as well as model creation.

DEEP LEARNING APPROACHES FOR TEXT RECOGNITION

Deep learning's objective approach is to solve the complexity of the sophisticated factors of the input through using high–level features. Deep learning technology is based on the idea that nothing fundamentally challenges systems to improve performance, e.g., machine handwriting recognition achieves human quality standards. Many types of structures and algorithms can be used to formulate the deep learning concept.

Some of deep learning's important approaches are:

Convolutional Neural Network (CNN)

CNN [6] is a method of deep learning algorithm that is specifically trained to perform with image files. A simple class that perfectly represents the image in CNN, processed through a series of convolutional layers, a pooling layer, and fully connected layers. CNN can learn multiple layers of feature representations of an image by applying different techniques. Low-level features such as edges and curves are examined by image classification in this method and a sequence of convolutional layers helps in building up to more abstract. CNN provides greater precision and improves performance because of its exclusive characteristics, such as local connectivity and parameter sharing. The input layer, multiple hidden layers (convolutional, normalization, pooling), and a fully connected and output layer make up the system of CNN. Neurons in one layer communicate with some neurons in the next layer, making the scaling simpler for higher resolutions.

In the input layer, the input file is recorded and collected. This layer contains information about the input image's height, width, and several channels (RGB information). To recognize the features, the network will use a sequence of convolutions and pooling operations in multiple hidden layers. Convolution is one of the most important components of a CNN. The numerical mixture of multiple functions to produce a new function is known as convolution. Convolution is applied to the input image via a filter or, to produce a feature map in the case of a CNN. The input layer contains $n \times n$ input neurons which are convoluted with the filter size of $m \times m$ and return output size of $(n - m + 1) \times (n - m + 1)$. On our input, we perform several convolutions, each with a different filter. As a result, different feature maps emerge. Finally, we combine these entire feature maps to create the convolution layer final output. To reduce the input feature space and hence reduces the higher computation; a pooling layer is placed between two convolutional layers. Pooling allows passing only the values you want to the next layer, leaving the unnecessary behind. This reduces training time, prevents overfitting, and helps in feature selection. The max-pooling operation takes the highest value from each sub-region of the image vector while keeping the most information, this operation is generally preferred in modern applications. CNN's architecture, like regular neural network architecture, includes an activation function to present non-linearity into the system. Among the various activation functions used extensively in deep learning models, the sigmoid function rectified linear unit (ReLu), and softmax are some well-known examples. In CNN architecture, the classification layer is the final layer. It's a fully connected feed-forward network that's most commonly used as a classifier. This layer determines predicted classes by categorizing the input image, which is accomplished by combining all the previous layers' features.

Image recognition, image classification, object detection, and face recognition are just a few of the applications for CNN. The most important section in CNN is the feature extraction section and classification section.

Recurrent Neural Network (RNN)

RNN is a deep learning technique that is both effective and robust, and it is one of the most promising methods currently in use because it is the only one with internal storage. RNN is useful when it is required to predict the next word of sequence [7]. When dealing with sequential data (financial data or the DNA sequence), recurrent neural networks are commonly used. The reason for this is that the model employs layers, which provide a short-term memory for the model. Using this memory, it can more accurately determine the next data and memorize all the information about what was calculated. If we want to use sequence matches in such data, we'll need a network with previous knowledge of the data. The output from the previous step is fed into the current step in this approach. The architecture of RNN includes three layers: input layer, hidden layer, and output layer. The hidden layer remembers information about sequences.

If compare RNN with a traditional feed-forward neural network(FNN), FNN cannot remember the sequence of data. Suppose we give a word "hello" as input to FNN, FNN processes it character by character. It has already forgotten about 'h' 'e' and 'l' by the time it gets to the character 'o'. Fortunately, because of its internal memory, a recurrent neural network can remember those characters. This is important because the data sequence comprises important information about what will happen next, that's why an RNN can perform tasks that some other techniques cannot.

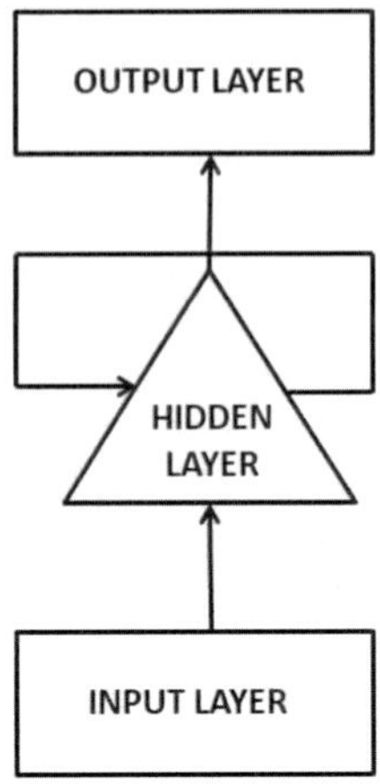

Figure 4. Recurrent Neural Network.

Application areas of RNN include sequence classification such as sentiment classification and video classification, etc. sequence labeling such as image captioning and named entry recognition, etc. and sequence generation such as machine translation, etc. Recurrent Neural Network is useful in time series prediction, and it has flexibility for handling various types of data.

Long Short Term Memory (LSTM)

LSTM is a difficult technique in deep learning to master. LSTM has feedback connections, unlike traditional feed-forward neural networks. It can process entire data sequences such as speech or video, as well as single data points such as images [8]. LSTM overcomes the problems of the RNN model. RNN model suffers from short-term memory. RNN model has no control over which part of the information needs to be carried forward and how many parts need to be forgotten. A memory unit called a cell is utilized by the LSTM which can maintain information for a sufficient period. LSTM networks are a type of RNN that can learn long chains of dependencies. LSTM has different memory blocks called cell which carries information throughout the processing of the sequence. The two states that are input to the next cell are the cell state and the hidden state. Three major techniques, referred to as gates, are used to manipulate this memory. A typical LSTM unit consists of a cell or memory block, an input gate, an output gate, and a forget-gate. The information in the cell is regulated by the three gates, and the cell remembers values for arbitrary periods. This model contains interacting layers in a repeating module.

Forget-gate layer is responsible for what to keep and what to throw from old information. Data that isn't needed in LSTM to comprehend the information of low significance is removed by multiplying a filter. This is mandated for the LSTM network's effectiveness to be optimized.

The input gate layer manages of determining what data should be stored in the cell state. To control what values should be assigned to the cell state, a sigmoid function is used. In the same way that the forget-gate filters all the data, this one does as well. The cell state is only updated with information that is both important and not useless.

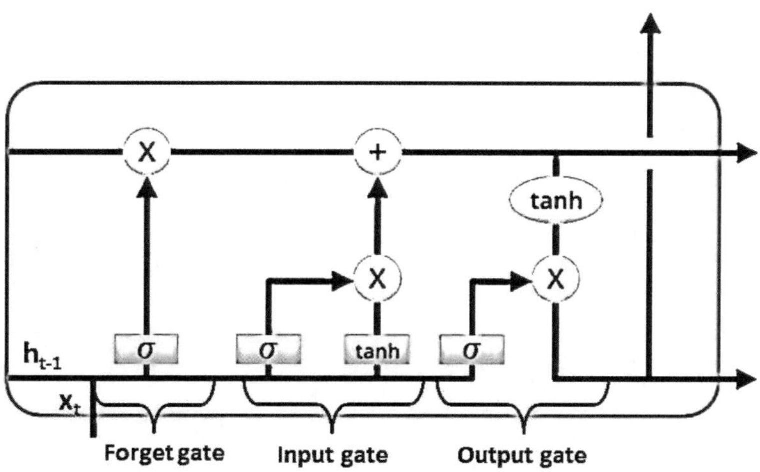

Figure 5. A typical LSTM architecture.

Summarized Table for Literature Review

Table 1. Table showing the different methodology and datasets used in different text recognition models, and the results achieved by the model

Reference No.	Methodology	Datasets	Results/Conclusion
[9]	Gated convolutions Encoder Interface The decoder is a BLSTM layer	IAM dataset RIMES	Line level: 3.2% CER(character error rate) 1.9% CER 7.9%WER
		IAM RIMES	Paragraph level: 3.3% CER 10.1% WER 2.2% CER 7.9% WER
[10]	BLSTM+CNN(GRCL) SGD	Synthetic from many historical documents (handwriting synthesis pipeline) + IAM dataset	WER = (10.8 to 30.1%) CER = (4.0 to 12.4%)
[11]	Multiscale CNN+BLSTM (Text localization +Text recognition) Adam	IAM dataset	CER = 6.4 to 28.3%
[12]	CNN + BLSTM + CTC (4 conv. + 3 stacked BLSTM)	Marriage records	Curriculum learning use of language models affect the accuracy
[13]	MDLSTM-RNN + CTC (No explicit line segmentation) The curriculum method could be useful.	IAM dataset	CER=6.8 to 8.4%
[14]	Manifold mixup (GAN and interpolating can be used) Glorot initialization of weights RMS Prop	Maurdor dataset(French) IAM dataset	9.39(without mixup) to 8.9 (with mixup)% 4.68% (without mixup) to 4.64% (with mixup) Manifold mixup is better

Reference No.	Methodology	Datasets	Results/Conclusion
[15]	Example-packing MDLSTM	IAM dataset	Speed of Training increases
[16]	Active learning is for curriculum. CNN + MDLSTM + CTC	IAM dataset RIMES and OpenHaRT	Dropped from 22 to 17% Slight improvement in both
[17]	CNN & LSTM decoder SGD and AdaDelta	IAM dataset	16.85 CER% (comparison table is given in this paper)
[18]	CNN-BLSTM-CTC (transfer learning, so we reduce the training set)	IAM dataset Washington Parzinal	Trained in lesser time. CER = 3.0 to 3.3%
[19]	Feature extractor module and sequence learning BLSTM encoder + LSTM decoder Layer normalization, Focal loss, Beam search	IAM dataset RIMES	8.1% CER16.7% WER(IAM) 3.5% CER 9.6% WER (RIMES) (Beat the current state-of-the-art at word level accuracy but not at character level accuracy)
[20]	CTC loss, RMS Prop, Dynamic augmentation, batch normalization	IAM dataset RIMES	Dropped from 8.2 to 6.3% CER Dropped from 3.3 to 2.6% CER WER also reduced
[21]	Encoder: Deep CNN + two BLSTM Decoder : LSTM Attention: 6 different types of attention mechanism is tried. Positional encoding	IAM dataset, ICFHR2016 READ (Bozen) dataset, StAZH dataset	4.66% CER
[22]	Metalabeler • Content based • Score based • Ranked based	Yahoo! Directory Multitopic web Categorization data	Micro-f1 and macro-f1 Improved by 1-9% at a different level concerning hierarchical models.
[23]	Text prediction Synthesis network (LSTM+RNN)	IAM online handwriting database	Given real-valued complex and long-range sequences.

Table 1. (Continued)

Reference No.	Methodology	Datasets	Results/Conclusion
[24]	DropOut and DropConnect methods, RNN	IAM RIMES OpenHaRT	It improved performance by 30-40%
[25]	Segmentation CNN	12 videos of French news broadcast programs	Character recognizer produces excellent results and outperforms methods based on SVM models by a significant margin.
[26]	MDLSTM(GPU-Based) CUDA and cuBLAS	IAM DATASET RIMES DATASET	On the development set, WER improved by 7.1%. On the evaluation set, WER improved by 9.3%
[27]	Recurrent encoder Decoder(bidirectional) Content-based attention mechanism	RIMES Dataset	WER=3.6% CER=1.3%
[28]	HCCR-GoogleNet	CASIA-HW (DB1.0) CASIA-HW (DB1.1)	Best testing error rate=3.26%
[29]	CNN(DCNN+RNN) ADADELTA	Synthetic dataset	When compared to traditional methods, CRNN achieves superior or desirable results.
[30]	TextBlock FCN Character-Centroid FCN	MSRA-TD500 dataset Incidental Scene Text dataset	Improvements on MSRA-TD500 dataset Precision =0.02, Recall = 0.04, and F-measure = 0.03.
[31]	Integrated Segmentation and Recognition	Different parts of India where the Hindi language is used	Accuracy of 89% on 10,000 words
[32]	HoG and LBP	Printed-word MILE Database	Accuracy over 11 scripts is 97.7%.
[33]	CNN-based method	JEITA-HP database IAM database	99.97% identification rate on the JEITA-HP database Accuracy=91.5% higher than handcrafted features
[34]	DCRN CNN+BLSTM+ CTC-decoder	TUAT Kondate database	Label error rate = 6.44%, Sequence error rate = 25.89%

Reference No.	Methodology	Datasets	Results/Conclusion
[35]	DCNN-based framework Hybrid serial-parallel (HSP) strategy	CASIA-OLHWDB1.0 CASIAOLHWDB1.1	Accuracy =97.20% Accuracy =96.87%
[36]	Contextual sub-characters HMMs Multi-stream contextual sub-character HMM	IFN/ENIT database	Recognition rate = 85.12%
[37]	Class-based contextual model	IFN/ENIT database	More compact
[38]	TextBoxes	SynthText ICDAR 2013 (IC13) Street View Text (SVT)	Both high accuracy and efficiency achieved
[39]	ASTER: rectification network and recognition network.	SynthText IIIT5k-Words Street View Text (SVT)	Cropped text recognition and end-to-end recognition
[40]	Discriminative Convolutional Neural Network (DisCNN) A deep learning-based method	SIW-13 dataset	Successfully applied on script identification, natural scene images, in documents and in videos
[41]	Multi-scale spatial partition network (MSP-Net) CNN-variant	TextDis benchmark ICDAR2003 dataset Hua's dataset	Provides effectiveness and robustness
[42]	Feature Enhancement Network (FEN) Text Detection Refinement	ICDAR 2011 ICDAR 2013	Improving the recall rate Accurately detecting text regions Solving the sample imbalance problem by positives mining strategy
[43]	Adaptive embedding gate (AEG) module ADADELTA	IIIT5K, SVT, SVT-P, CUTE80, and ICDAR datasets.	Boost recognition performance and bring better robustness.
[44]	TextSnake Fully Convolutional Network (FCN)	TotalText CTW1500 SynthText MSRA-TD500	Outperforms, by more than 40% in F-measure on Total-Text.
[45]	Hybrid CRNN CNN+RNN	YouTube News dataset and EU Speech Repository etc.	Successfully applied to a wide variety of languages.

Table 1. (Continued)

Reference No.	Methodology	Datasets	Results/Conclusion
[46]	A scene text image synthesis engine that renders images with 3D graphics engines	VISD SynthText3D	Effectiveness in scene text detection and recognition models
[47]	Semi-supervised neural networks. DNN	SVHN dataset FSNS dataset	Achieve competitive results, but still not fully capable of detecting text in arbitrary locations in the images.
[48]	A ResNet based feature extractor Spatial Transformer RAdam optimizer	ICDAR 2013(IC13), IIIT5K, SynthText, SVTP, CUTE80	Compared to other approaches, achieved state-of-the-art results.
[49]	Bag of Feature-based technique Sparse neural networks models	Chars74K and ICDAR2003 ARASTI	A reliable recognition method for cropped English and Arabic characters.
[50]	A deep Sparse Auto-encoder (SAE)-based strategy	Chars74K and ICDAR03-CH ARASTI	Accuracy=88.5%
[51]	Connectionist Temporal Classification (CTC) Convolutional feature map	IIIT-5K, SVT, ICDAR03/13 and TRW15 datasets	Achieves superior performance. Highly parallel and enables fast recognition.
[52]	Sliding window and sequence learning CNN	IIIT5k, SVT and ICDAR 2003/2013 datasets	Improvement in performance and interpretability.
[53]	Deep convolutional feature Representation CNN	IAM Handwriting Database	WER=6.69% CER=3.72%
[54]	End2End embedding framework CRNN BLSTM	IAM handwritten dataset	WER=5.10% CER=2.66%
[55]	CNN-RNN hybrid architecture	IAM RIMES	Achieved effectiveness and improved recognition rates
[56]	Synthetic Word Image Rendering Data augmentation schemes	IAM handwriting dataset IIIT-HWS dataset	Renders large scale synthetic data
[57]	Deep embeddings representation	Hindi documents	Improvement in word recognition rate

The output gate determines the selection and displaying valuable data about the current state of the cell. The cell is first used to generate a vector using the tanh function. The data is then filtered using the sigmoid function

LSTMs hold great promise for solving problems that involve sequences and time series.

CONCLUSION

Due to the variety of writing styles used in multiple languages, text recognition is a difficult task. In this chapter, we have looked at how text recognition works and how different steps like segmentation, feature extraction, and classification are used. Deep learning approaches have also been explained, which are beneficial in text recognition. The analysis in this chapter revealed that there is still scope for improvement in both the algorithms and the word recognition rate.

REFERENCES

[1] Polaiah, B., Naga, s., Gautham, K. P. & Lalitha Rao Sharma Polavarapu, S. D. (2019). Handwritten Text Recognition using Machine Learning Techniques in Application of NLP. *(IJITEE)*. ISSN: 2278-3075, Volume *9*, Issue-2.

[2] Yogesh Kumar, D. & Pulkit Jain. (2018). Comprehensive Survey on Machine Learning Application for Handwriting Recognition. *International Journal of Applied Engineering Research*, ISSN 0973-4562 Volume *13*, Number 8 pp. 5823-5830.

[3] Singh, K. R. (2015). A Technical Review on Text Recognition from Images *IEEE Sponsored 9th International Conference on Intelligent Systems and Control (ISCO).*

[4] Ayush Purohit. & Shardul Singh Chauhan. (2016). A Literature Survey on Handwritten Character Recognition", *IJCSIT*, Vol. *7*(1).

[5] Manoj Sonkusare. & Narendra Sahu. (2016). A Survey On Handwritten Character Recognition (HCR) Techniques For English Alphabets. *Advances in Vision Computing: An International Journal (AVC)*, Vol. *3*, No.1.

[6] Nal Kalchbrenner, Edward Grefenstette. & Phil Blunsom. (2014). A convolutional neural network for modelling sentences. In *Proceedings of ACL.*

[7] Junyoung, C., Gulcehre, C., KyungHyun Cho. & Yoshua Bengio. (2014). Empirical evaluation of gated recurrent neural networks on sequence modeling. arXiv preprint *arXiv:1412.3555.*

[8] Hochreiter, S. & Schmidhuber, J. (1997). Long short-term memory," *Neural Computation*, vol. *9*, no. 8, pp. 1735–1780.

[9] Bluche, T. & Messina, R. (2017, November). Gated convolutional recurrent neural networks for multilingual handwriting recognition. In *2017 14th IAPR International Conference on Document Analysis and Recognition (ICDAR)*, (Vol. *1*, pp. 646-651). IEEE.

[10] Ingle, R. R., Fujii, Y., Deselaers, T., Baccash, J. & Popat, A. C. (2019). A Scalable Handwritten Text Recognition System. arXiv preprint *arXiv:1904.09150.*

[11] Chung, J. & Delteil, T. (2019, September). A Computationally Efficient Pipeline Approach to Full Page Offline Handwritten Text Recognition. In *2019 International Conference on Document Analysis and Recognition Workshops (ICDARW),* (Vol. *5*, pp. 35-40). IEEE.

[12] Carbonell, M., Villegas, M., Fornés, A. & Lladós, J. (2018, April). Joint recognition of handwritten text and named entities with a neural end-to-end model. In *2018 13th IAPR International Workshop on Document Analysis Systems (DAS),* (pp. 399-404). IEEE.

[13] Bluche, T. (2016). Joint line segmentation and transcription for end-to-end handwritten paragraph recognition. In *Advances in Neural Information Processing Systems*, (pp. 838-846).

[14] Moysset, B. & Messina, R. (2019). Manifold Mixup improves text recognition with CTC loss. arXiv preprint *arXiv:1903.04246.*

[15] Wenniger, G. M. D. B., Schomaker, L. & Way, A. (2019). No Padding Please: Efficient Neural Handwriting Recognition. arXiv preprint *arXiv:1902.11208.*

[16] Louradour, J. & Kermorvant, C. (2014, April). Curriculum learning for handwritten text line recognition. In *2014 11th IAPR International Workshop on Document Analysis Systems,* (pp. 56-60). IEEE.

[17] Poulos, J. & Valle, R. (2019). *Character-Based Handwritten Text Transcription with Attention Networks.*

[18] Aradillas, J. C., Murillo-Fuentes, J. J. & Olmos, P. M. (2018). Boosting Handwriting Text Recognition in Small Databases with Transfer Learning. arXiv preprint *arXiv:1804.01527.*

[19] Chowdhury, A. & Vig, L. (2018). An efficient end-to-end neural model for handwritten text recognition. arXiv preprint *arXiv: 1807.07965.*

[20] Puigcerver, J. (2017). Are multidimensional recurrent layers really necessary for handwritten text recognition? In *2017 14th IAPR International Conference on Document Analysis and Recognition (ICDAR)*, (Vol. 1, pp. 67-72). IEEE.

[21] Michael, J., Labahn, R., Grüning, T. & Zöllner, J. (2019). Evaluating Sequence-to-Sequence Models for Handwritten Text Recognition. arXiv preprint *arXiv:1903.07377.*

[22] Lei Tang, Suju, R. & Vijay, K. N. (2009). *Large Scale Multi-Label Classification via MetaLabeler.*

[23] Alex Graves. (2014). *Generating Sequences With Recurrent Neural Networks*. 1308.0850v5 [cs.NE].

[24] Vu Pham, T. Bluche, Christopher, K. & Louradour, J. (2014). Dropout improves Recurrent Neural Networks for Handwriting Recognition. *arXiv:1312.4569v2* [cs.CV].

[25] Khaoula, E., Garcia, C. & Pascale, S. (2011). *Comprehensive Neural- Based Approach for Text Recognition in Videos using Natural Language Processing*. ICMR.

[26] Voigtlaender, P., Doetsch, P. & Ney, H. (2016). *Handwriting recognition with large multidimensional long short-term memory recurrent neural networks*. ICFHR.

[27] Doetsch, P., Zeyer, A. & Ney, H. (2016). Bidirectional decoder networks for attention-based end-to-end offline handwriting recognition," *International Conference on Frontiers in Handwriting Recognition*, pp. 361–366.

[28] Zhuoyao, Z., Lianwen, J. & Zecheng Xie. (2015). High Performance Offline Handwritten Chinese Character Recognition Using GoogLeNet and Directional Feature Maps. *ICDAR*, (pp. 846-850). "

[29] Baoguang Shi, Xiang Bai. & Cong Yao. (2016). *An End-to-End Trainable Neural Network for Image-based Sequence Recognition and Its Application to Scene Text Recognition*. IEEE (TPAMI).

[30] Zheng Zhang, Chengquan Zhang, Wei Shen, Cong Yao, Wenyu Liu, & Xiang Bai. (2016), "*Multi-Oriented Text Detection with Fully Convolutional Networks*". IEEE.

[31] Swapnil Belhe, Chetan P., Akash D., Saumya J. & Kapil M. (2016). Hindi Handwritten Word Recognition using HMM and Symbol Tree. *Workshop on Document Analysis and Recognition*.

[32] Saumya, J., Kapil Mehrotra, Atish Vaze. & Swapnil Belhe. (2014). Multi-script Identication from Printed Words. *International Conference Image Analysis and Recognition*.

[33] Hung, T. N., Cuong, T. N., Takeya Ino, Bipin, I. & Masaki, N. (2019). Text-Independent Writer Identification using Convolutional Neural Networks. *Pattern Recognition Letters*, (vol. *121*).

[34] Nam-Tuan Ly, Cuong-Tuan Nguyen, Kha-Cong Nguyen. & Masaki N. (2017). Deep Convolutional Recurrent Network for Segmentation-free Offline Handwritten Japanese Text Recognition. *IAPR (ICDAR),* (vol. 7).

[35] Weixin, Y., Lianwen, J., Zecheng, Xie. & Ziyong Feng. (2015). Improved Deep Convolutional Neural Network For Online Handwritten Chinese Character Recognition using Domain-Specific Knowledge. *(ICDAR), IEEE.*

[36] Irfan Ahmad, Gernot A. Fink. & Sabri A. Mahmoud. (2014). Improvements in Sub-Character HMM Model Based Arabic Text Recognition. *International Conference on Frontiers in Handwriting Recognition*. IEEE.

[37] Irfan Ahmad. & Gernot A Fink. (2016). Class-Based Contextual Modeling for Handwritten Arabic Text Recognition. *(ICFHR), IEEE*.

[38] Minghui, L., Baoguang, S., Xiang, Bai., Xinggang, W. & Wenyu, L. (2017). "TextBoxes: A Fast Text Detector with a Single Deep Neural Network", *AAAI Conference on Artificial Intelligence*.

[39] Baoguang, S., Mingkun, Y., Xinggang, W., Pengyuan, L., Yao, C. & Xiang, B. (2018). "*ASTER: An Attentional Scene Text Recognizer with Flexible Rectification*", IEEE.

[40] Baoguang Shi, XiangBai n. & CongYao. (2016). Script identification in the wild via discriminative convolutional neural network. *Pattern Recognition*, (vol. *52*).

[41] Xiang Baia, Baoguang Shia, Chengquan Zhanga, Xuan Caib. & Li Qib. (2017). Text/non-text image classification in the wild with convolutional neural Networks. *Pattern Recognition*, (vol. *66*).

[42] Sheng Zhang, Yuliang Liu, Lianwen Jin. & Canjie Luo. (2018). Feature Enhancement Network: A Refined Scene Text Detector. *AAAI Conference on Artificial Intelligence*, (vol. *32*).

[43] Xiaoxue Chen, Tianwei Wang, Yuanzhi Zhu, Lianwen Jin. & Canjie Luo. (2020). Adaptive embedding gate for attention-based scene text recognition. *Neurocomputing*, (vol. *381*).

[44] Shangbang Long, Jiaqiang Ruan, Wenjie Zhang, Xin He, Wenhao Wu. & Cong Yao. (2018). *TextSnake: A Flexible Representation for Detecting Text of Arbitrary Shapes*.

[45] Christian Bartz, Tom Herold, Haojin Yang. & Christoph Meinel. (2017). Language Identification Using Deep Convolutional Recurrent Neural Networks. *International conference on neural information processing*, Springer.

[46] Shangbang Long. & Cong Yao. (2020). UnrealText: Synthesizing Realistic Scene Text Images from the UnrealWorld. *Proceedings of*

the IEEE/CVF Conference on Computer Vision and Pattern Recognition, (pages. 5488-5497).

[47] Christian Bartz, Haojin Yang. & Christoph Meinel (2018). SEE: Towards Semi-Supervised End-to-End Scene Text Recognition. *Proceedings of the AAAI Conference on Artificial Intelligence*.

[48] Christian Bartz, Joseph Bethge, Haojin Yang. & Christoph Meinel. (2019). KISS: Keeping It Simple for Scene Text Recognition. arXiv preprint *arXiv:1911.08400*.

[49] Maroua Tounsi, Ikram Moalla. & Adel M Alimi. (2016). Supervised Dictionary Learning in BoF Framework for Scene Character Recognition. *23rd International Conference on Pattern Recognition (ICPR).*

[50] Maroua Tounsi, Ikram Moalla, Frank Lebourgeois. & Adel M Alimi. (2018). *Multilingual Scene Character Recognition System using Sparse Auto-Encoder for Efficient Local Features Representation in Bag of Features*.

[51] Fei Yin, Yi-Chao Wu, Xu-Yao Zhang. & Cheng-Lin Liu (2017). *Scene Text Recognition with Sliding Convolutional Character Models*.

[52] Yi-Chao Wu, Fei Yin, Xu-Yao Zhang, Li Liu. & Cheng-Lin Liu. (2018). SCAN: Sliding Convolutional Attention Network for Scene Text Recognition. arXiv preprint *arXiv:1806.00578*.

[53] Praveen Krishnan, Kartik Dutta. & Jawahar, C. V. (2016). Deep feature Embedding for Accurate Recognition and Retrieval of Handwritten Text. *International Conference on Frontiers in Handwriting Recognition*.

[54] Praveen Krishnan, Kartik Dutta. & Jawahar, C. V. (2018). Word Spotting and Recognition using Deep Embedding. *Document Analysis Systems*.

[55] Kartik Dutta, Praveen Krishnan, Minesh Mathew. & Jawahar, C. V. (2018). Improving CNN-RNN Hybrid Networks for Handwriting Recognition. *International Conference on Frontiers in Handwriting Recognition.*

[56] Praveen Krishnan. & Jawahar, C. V. (2016). *Generating Synthetic Data for Text Recognition.*

[57] Siddhant Bansal, Praveen Krishnan. & Jawahar, C. V. (2020). Fused Text Recogniser and Deep Embeddings Improve Word Recognition and Retrieval. *International Workshop on Document Analysis Systems*. Springer, Cham, (pp. 309-323).

In: Deep Learning and its Applications ISBN: 978-1-68507-185-1
Editor: Arvind Kumar Tiwari

Chapter 3

APPLICATIONS OF DEEP LEARNING IN DIABETIC RETINOPATHY DETECTION

Imtiyaz Ahmad*, Vibhav Prakash Singh† and Suneeta Agarwal‡
Department of Computer Science and Engineering,
Motilal Nehru National Institute of Technology Allahabad,
Prayagraj, UP, India

ABSTRACT

Diabetic Retinopathy (DR) is one of the common issues of diabetic Mellitus that affects the eyesight of humans by causing lesions in their retinas. DR is mainly caused by the damage of blood vessels in the tissue of the retina, and it is one of the leading causes of visual impairment globally. It can even cause blindness if not detected in its early stages. To reduce the risk of eyesight loss, early detection and treatment are pretty necessary. The manual process by ophthalmologists in detection DR requires much effort and time and is costly also. Many

* Corresponding Author's Email: amaanimtiyaz.ai@gmail.com.
† Corresponding Author's Email: vibhav@mnnit.ac.in.
‡ Corresponding Author's Email: suneeta@mnnit.ac.in.

computer-based techniques reduce the manual effort, and deep learning is used more commonly in medical imaging. This chapter will discuss deep learning and how it is helpful in the early detection and classification of DR by reviewing some latest state-of-art methods. There are various datasets of colour fundus images available publically, and we have reviewed those databases in this chapter.

Keywords: diabetic retinopathy, fundus images, deep learning

INTRODUCTION

The terms Artificial Intelligence (AI), Machine Learning (ML), and Deep Learning (DL) are usually used identically but are not the same. AI is called a vast field of research where the goal is to make the device interact with its nature as an intelligent person. Machine Learning (ML) is a subset of AI where a machine learns to perform a task without explicit programming. Deep learning (DL) is a subset or sub-field of ML that deals with algorithms that use deep neural networks.

DL (also called hierarchical learning or deep structured learning) is a part of machine learning that is based on some set of algorithms, which performs a high level of abstractions in data [1-4]. Such algorithms develop a layered and hierarchical architecture of learning, understanding, and representing the data. This advanced learning technology is inspired by artificial intelligence, which imitates the deep, layered learning process of the human brain, which automatically extracts features and releases primary data [5, 6]. DL algorithms are useful as they can deal with large amounts of unsupervised data and naturally learn data representation in a deep layer-wise method which a simple ML algorithm can't do [7, 8].

Applications of DL in today's world are a comprehensive concept. Many deep learning architectures like Deep Neural Network (DNN), Convolutional Neural Network (CNN), and Recurrent Neural Network (RNN) have been implemented in areas like computer vision, speech recognition, NLP (Natural Language Processing), audio recognition and bioinformatics, etc. [9]. Deep learning can be depicted as a class of machine learning algorithms that uses a cascade of multiple layers for feature extraction and transformation. Output from the previous layer is used as input for the next layer. Algorithms of deep learning can be both supervised or unsupervised [9].

The number of parameterized transformations is a signal encounter as it propagates from the input layer to the output layer and the number of hidden layers present in the network. In deep networks, processing units with trainable parameters, like weights and thresholds, are the parameterized transformations. Figure 1 shows the difference between these two networks. A chain of transformations between the input and output layers is the credit assignment path (CAP), which may vary in length and defines connections between input and output.

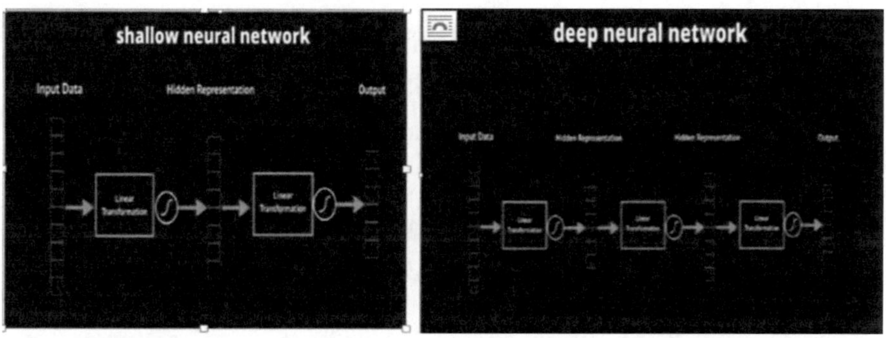

(a) (b)

Figure 1. (a) Simple NN and (b) Deep NN [10].

Deep Learning in the Detection of Diabetic Retinopathy

A few years back, we would have never thought of deep learning applications to develop virtual assistants (like Alexa, Siri, and Google Assistant) and self-driving cars. But now, these developments are part of our daily lives. Deep learning is fascinating to us with its ongoing actions, such as fraud detection and pixel restoration.

Although deep learning is achieving quite impressive results in real-world applications, it is essential to note that it is not magic to achieve those results; large amounts of data are required. Furthermore, learning from this amount of data is a very time-consuming and computationally demanding process.

Nevertheless, it is terrible how these algorithms can "learn" without telling the model what to look for - it learns based on experience and the examples given. And as mentioned, progress in this area has been made to develop surprising and valuable applications that we will discuss next.

Deep Learning is a very vast field to discuss and do research on. There are hundreds of applications present at the moment that use deep learning methods. Hundreds of fascinating applications will come in the future, like MIT is working on future prediction using deep learning methods.

Deep learning is widely used in the medical field—computer vision techniques like image segmentation, image classification, etc. Using different deep learning architectures (such as CNN, RNN, LSTM) can detect any disease from other image datasets. Deep learning helps medical experts diagnose disease more accurately with minimum error and allows them to treat it better, thus leading to better decisions.

This chapter will discuss eye disease, which can cause blindness known as Diabetic Retinopathy (DR). Early detection of DR is a critical task, and deep learning helps in its early detection.

Diabetic Retinopathy (DR)

Diabetes is a condition that occurs in the human body when the pancreas fails to produce the required insulin or when the body fails to process it properly. As it advances, it starts affecting the circulatory system of the human body, including the retina. It causes damage to the retinal blood vessels, leading to diabetic retinopathy by decreasing the patient's vision. This disease can cause permanent blindness to the affected person if appropriate treatment is not provided in the early stages.

The abnormal shift in blood sugar level starts happening in diabetes mellitus. Generally, glucose is converted into energy in the human body that helps to perform normal human functions. But in the worst-case scenario, there is an abnormal blood sugar level, and the excess blood sugar causes hyperglycemia. Non-proliferative Diabetic Retinopathy (NPDR) and Proliferative Diabetic Retinopathy (PDR) are two main stages of DR, as shown in Figure 2. NPDR is a condition in which the retina becomes inflamed (a case of macular edema) because of the accumulation of glucose that leads to leakage of blood vessels in the eyes. In a severe condition, retinal vessels might get blocked completely, which causes macular ischemia. There are different levels in NPDR in which sometimes the patient suffers from blurred vision or loses sight partially or entirely. PDR occurs in the advanced stage of diabetes, in which extra blood vessels start growing in the retina (a case of neovascularization). These new blood vessels are very narrow and brittle, tend to cause haemorrhages, and lead to partial or complete loss of vision.

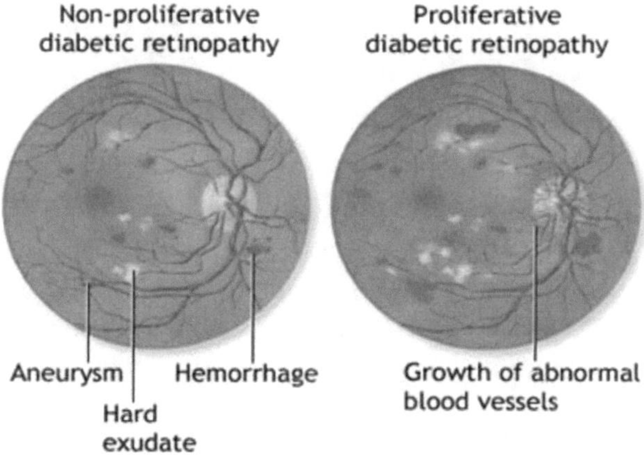

Figure 2. Figure showing an NPDR retina and PDR retina.

Automatic testing is essential for reducing manual effort. Making it expensive to detect deviations from retina images should be done automatically through digital photography techniques and imaging techniques. In DR the arteries that help in nourishing the retina start to leak fluid and blood into the retina causing visual aids known as lesions such as microaneurysms, hemorrhages, hard exudates, and cotton wool spots. These various Lesions are shown in Figure 2 and are described as under [10]:

- *Microaneurysms-* appear as a red dot and are evagination of capillaries of the retina. It can develop in any condition, which can cause retinal microvasculopathy. They have increased penetration and may bleed or leak, causing localized retinal haemorrhage or edema.
- *Hard Exudates-* are well restricted, white deposit or cream within the retina. They show the accumulation of fluid in the retina and are considered a threat to vision if it appears near the centre of the retina, i.e., Macula. They are usually seen in combination with microaneurysms.
- *Haemorrhages-* can be of a different shape and size depending on their location inside the retina. Dot haemorrhages are the most common DR haemorrhages. They are originated from the external capillary network of the retina and are small and round.
- *Cotton wool spots-* have white/yellow retinal lesions with distinct borders of feathers-represent oedema areas within the retinal nerve fiber layer due to focal ischemia. Usually, they get resolve within three months automatically. New lesions may appear in different areas if the underlying ischemic condition continues. They are usually associated with haemorrhage in the face and microaneurysms and represent retinal microvasculopathy.

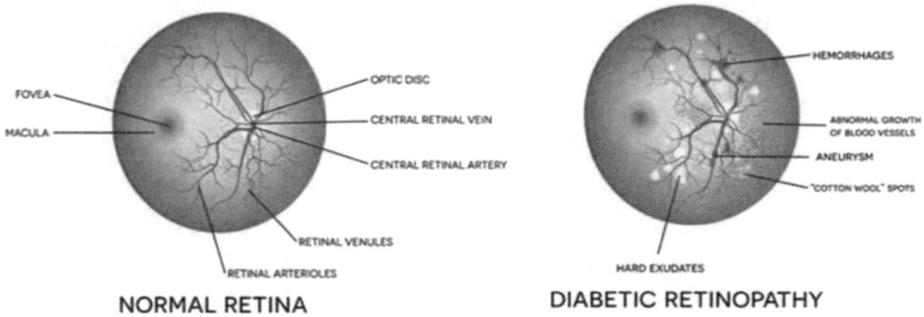

Figure 3. Various lesions of the retina and a normal retina.

Severity Levels of DR

Early Treatment Diabetic Retinopathy Study Research Group (ETDRS) and International Clinical Diabetic Retinopathy (ICDR) [12] has given different levels of severity of DR defined as under:

- *Level 0-* No retinopathy.
- *Level 1- Mild NPDR-* the presence of at least one micro-aneurysm with or without other lesions.
- *Level 2- Moderate NPDR-* the presence of many micro-aneurysms and retinal haemorrhages with or without cotton wool spots.
- *Level 3- Severe NPDR–* the presence of many haemorrhages and micro-aneurysms in four quadrants of the retina, cotton wool spots in two or more quadrants and Intra-retinal microvascular abnormalities in one or more quadrants.
- *Level 4- PDR-* it is an advanced stage of DR where new narrow and brittle or weak blood vessels are present with a high risk of leakage, and it can cause severe vision loss and sometimes even blindness. Figure 4 shows images of different levels of severity in DR.

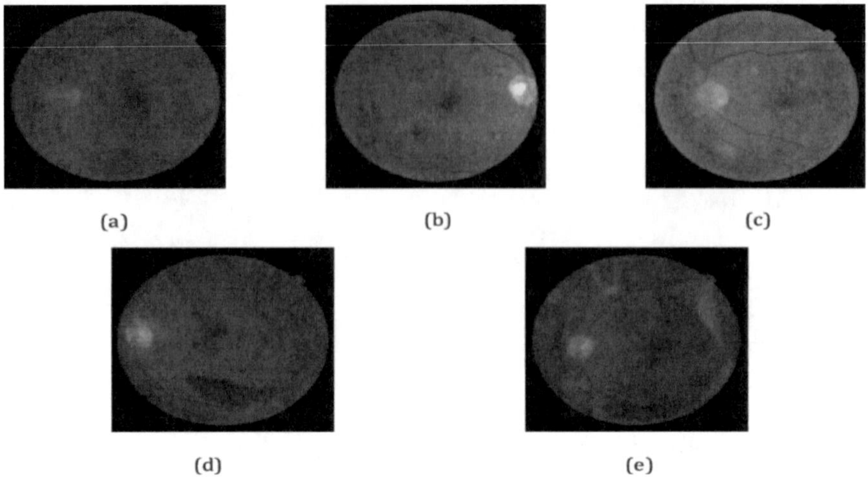

Figure 4. DR with increasing level of severity: (a) no DR; (b) mild NPDR; (c) moderate NPDR; (d) severe NPDR; (e) PDR.

Metrics for Evaluation

Following metrics are used for evaluation:

Accuracy: It is defined as the percentage of correct predictions made by a classifier compared to the actual value of the label. It can also be defined as the average number of correct tests in all tests [13]. To calculate accuracy, we use the equation:

$$Accuracy = (TN + TP)/(TN + TP + FN + FP). \quad (1)$$

Here, TP, TN, FP, and FN mean true positives, false negatives, false positives, and false negatives. True positive is a condition where if the class label of a record in a dataset is positive, the classifier predicts the same for that particular record. Similarly, a true negative is a condition where if the class label of a record in a dataset is negative, the classifier predicts the same for that particular record. False-positive is a condition where the class label of a record in a dataset is negative, but the classifier predicts the class label as positive. Similarly, a false negative is a condition where the class label of a record in a dataset is positive. Still, the classifier predicts the class label as negative for that record [13].

Sensitivity: It is defined as the percentage of true positives identified by the classifier while testing. To calculate it, we use the equation:

$$Sensitivity = (TP)/(TP + FN). \tag{2}$$

Specificity: It is defined as the percentage of true negatives which are rightly identified by the classifier during testing. To calculate it we use the equation:

$$Specificity = (TN)/(TN + FP). \tag{3}$$

Databases Available

Although several databases of fundus images are available publicly, the creation of quality retinal image databases is still in progress to train deep neural networks.

- *DRIVE [14] (Digital Retinal Image for vessel Extraction)* – This database contains 40 images collected from 400 samples of age 25 to 90 in the Netherland. Out of 40, 7 shows mild DR, whereas others are normal. Each set, i.e., training and testing, includes 20 images of different patients. For every image, manual segmentation known as truths or gold standards of blood vessels is provided.
- *STARE [15] (Structured Analysis of Retina)* – This database contains 20 retinal fundus images taken using a fundus camera. Datasets are divided into two classes or categories, one contains normal images, and the other includes images with various lesions.

- *CHASE [16]* – contains 28 images of 1280 * 960 pixels, taken from multi-ethnic children in England.
- *Messidor and Messidor-2 [17]* - These databases contain 1200 and 1748 images of the retina, respectively, taken from both eyes. Messidor-2 is an extension of the Messidor database taken from 874 samples.
- *EyePACS-1 [18]* - This database contains macula-centred images of 9963 subjects taken from different cameras in May-October 2015 at EyePACS screening sites.
- *APTOS [19] (Asia Pacific Tele-Ophthalmology Society)* – contains 3662 training images and 1928 testing images. Images are available with the ground truths classified based on severity of DR rating on a scale of 0 to 4.
- *Kaggle [20]* - contains 88,702 images of the retina with different resolutions and are classified into 5 DR stages. Many images are of bad quality, and also some of the ground truths have incorrect labelling.
- *IDRID [21] (Indian Diabetic Retinopathy Image Dataset)* – contains 516 retinal fundus images captured by a retinal specialist at an Eye Clinic located in Nanded, Maharashtra, India.
- *DIARETDB1 [22]* contains 89 retinal fundus images of size 1500* 1152 pixels, including 5 normal images and all other 84 DR images.
- *DDR [23] (Diagnosis of Diabetic Retinopathy)* – contains 13,637 retinal fundus images showing five stages of DR. From the dataset, 757 images show DR lesions.
- Others like E-ophtha [24], HRF [25], ROC [26], and DR2 [27] etc.

Process of Detection of DR Using Deep Learning

There are various numbers of supervised learning methods and unsupervised learning methods available for detecting Diabetic

Retinopathy. Deep learning is one technique widely used in medical imaging applications like image classification, image segmentation, image retrieval, image detection, and registration of images. For the detection and classification of diabetic retinopathy, Deep Learning techniques or deep neural networks have been widely used. Deep neural networks produce outstanding results in the removal of default features and isolation. Unlike machine learning methods, the performance of deep learning methods increases with an increase in the number of training datasets because of an increase in learned features. There is a various number of deep neural networks like CNN (Convolutional Neural Network), RNN (Recurrent Neural Network), LSTM (Long Short Term Memory), GRU (Gated Recurrent Unit), Autoencoders, RBM (Restricted Boltzmann Machine), DBN (Deep Belief Network), DSN (Deep Stacking Network), Self-Organizing Maps, etc. Still, CNN has been widely used in medical imaging and is highly effective [11]. Deep networks are much more powerful by using strategies such as the dropout function that helps the network produce relevant results even when few features are missing in the test dataset. In addition, ReLUs (Direct Line Units) function is used as a transfer function in CNNs that helps in effective training as they do not disappear too much like the sigmoid function and tangent function used by standard ANNs. The basic architecture of CNN is that it works in different layers like Convolutional layers, pooling layers, fully connected layers, Dropout, and Activation function at last. In the convolution layer, different types of filters are used to extract features from the image. The subsampling (or pooling) layer acts as feature selection and makes the network potent to changes in size and orientation of the image. Average pooling and max pooling are mostly used in the pooling layer. A fully connected layer is used to define the whole input image. Several pre-trained CNN architectures are present at the moment on ImageNet, such as LeNet, AlexNet, VGG, ResNet, GoogleNet, and more.

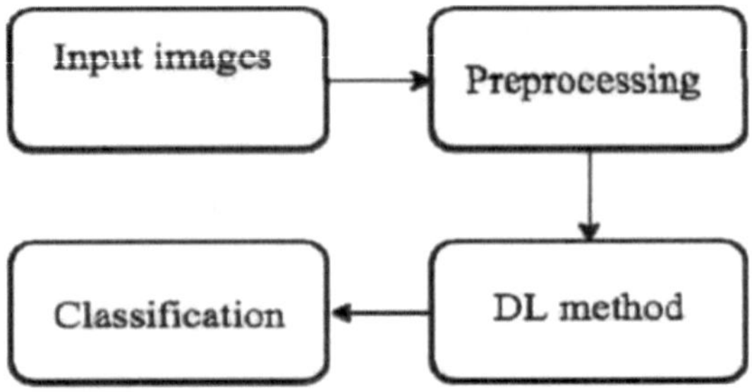

Figure 5. Process of classification of DR image by using DL method [11].

The detection and classification of DR using ML generally starts with the collection of datasets and then applying some image processing techniques to enhance and improve the quality and information of the images. After that, take the image as input, use the DL method and extract features, and finally do classification [11]. The steps are shown in Figure 5.

DR Screening Systems

Many researchers have tried to automate the detection and classification of DR lesions using Deep Learning. The classification methods used can be categorized into 4 parts i.e., binary classification, multi-level classification, vessel-based, and lesion-based classification. Methods and results used by different researchers for the detection of DR are summarized in Table 1.

Table 1. Methods used for classification and detection of DR

Author's Reference	DL method used	Database used	AUC	Accuracy (%)	Sensitivity (%)	Specificity (%)
[28]	CNN	Kaggle	-	94.5	-	-
[31]	CNN	Kaggle	-	75	30	95
[33]	CNN (Inception-v3)	Messidor2	-	-	96	93.9
[32]	CNN	Messidor2	0.98	-	96.8	87
[38]	CNN	DIARETDB	-	99.17	100	98.41

Author's Reference	DL method used	Database used	AUC	Accuracy (%)	Sensitivity (%)	Specificity (%)
[29]	CNN(ResNet34)	Kaggle	-	85	86	-
[30]	CNN	Messidor2, DR2, Kaggle	98.2%,98%, -	-	-	-
[35]	CNN(AlexNet, VggNet)	Messidor	-	98.15	98.940	97.87
[34]	CNN(AlexNet)	Messidor	-	96.35	92.35	97.45
[41]	CNN(ResNet)	DRIVE	0.973	95.10	79.3	97.4
[36]	CNN	Their dataset and Messidor	- and 0.97	92.65 and -	99.39 and 92.6	99.93 and 96.20
[43]	CNN	DRIVE, CHASE	0.956, 0.957	95.8, 96.0	86.4, 87.78	96.65, 96.8
[42]	CNN	DRIVE, STARE, CHASE	98.3%, 98.7%, 98.9%	95.8, 96.7, 96.8	79.97, 80, -	98.13, 98.6, 98.8
[39]	CNN(LeNet, Unet)	DIARETDB1	0.4823	-	48.71	-
[40]	DRN	E-optha	0.964	-	0.922	-

Binary Classification

In binary classification, DR images are classified into two classes, i.e., Yes and No. Yes represents infected images or images with the presence of DR, and No means images without DR. K. Xu et al. [28] used Kaggle [20] dataset (used 1000 images) and classified images into 2 classes (Normal and DR images). At first, they performed data augmentation by applying several images to increase the dataset and resizing images to 224*224*3. Then they put those images as an input to CNN architecture and got an accuracy of 94.5%. Their architecture included 8 Convolutional layers, four max-pooling layers, and two fully connected layers. For the classification, they used the SoftMax function at the last layer of CNN.

MT Esfahan et al. [29] worked on Kaggle [20] dataset (used 35000 images) and classified images into two classes (Normal and DR images) by using ResNet34 (a pre-trained CNN architecture). At first, they applied some image processing techniques like Gaussian filter and image normalization to increase the quality of images. Then they put those images as input to ResNet34 and got accuracy and sensitivity of 85% and 86%, respectively.

R. Pires et al. [30] used custom CNN architecture consists of 16 layers. They used Kaggle [20] dataset (used 88702 images) for training and testing. Also, they have used Messidor2 and DR2 datasets (1748 and 520 images, respectively). They classified images into two classes. During training, multi-image resolution and two-fold cross-validation were used. To reduce the overfitting problem, dropout and L2 regularization were used. They got an AUC of 98.2% on testing the Messidor2 dataset.

Multi-Level Classification

In Multi-level classification, DR images are classified into many classes. H. Pratt et al. [31] classified datasets from Kaggle [20] into 5 stages using a method based on CNN. At first, they performed image resizing (512 * 512) and colour normalization. Then they put those images as input to their customized CNN architecture consisting of ten, eight, and three convolutional, max-pooling, and fully connected layers. For classification of 80000 test images, the SoftMax function is used, and to reduce the overfitting problem, dropout and L2 regularization were used. Their results had accuracy, sensitivity, and specificity of 75%, 30%, and 95%, respectively. Unfortunately, only one dataset was used to evaluate their CNN architecture because their architecture failed to detect lesions in the images.

Gulshan et al. [33] used Messidor2 [17] and eyepacs1 [18] datasets (used 1748 and 9963 images, respectively) to detect DR and DME (diabetic macular edema). At first, they normalized and resized (to 299 pixels) to images and fed them to CNN as input. They used pre-trained InceptionV3 architecture and trained about 10 CNNs with different numbers of images. Images were classified into other classes like referable DME, moderate DR, severe DR, and fully gradable and got a specificity of 93% and sensitivity of 96% in Messidor2 dataset and 97.5 in eyepacs1 dataset.

M. Abramoff et al. [32] combined the CNN with IDX-DR device for detection and classification of DR. They used Messidor2 [17] dataset (used 1748 images) and applied data augmentation on it. They integrated many CNNs using Random Forest classifier. Images were classified into three classes (no DR, referable DR, and vision-threatening DR). They got AUC, specificity, the sensitivity of 0.980, 87%, and 96.8%. They did not consider the five stages of DR; instead, the Mild DR stage was considered no DR.

T. Shathi et al. [34] used Messidor [17] dataset (1190 images) and detected stages of DR using pre-trained CNN architecture, AlexNet. At first, resizing was done, and a green channel was extracted. Then those images were used as input into CNN and achieved 96.35% accuracy. The architecture does not detect lesions of infected images.

M. Rehman et al. [35] used Messidor datasets (1200 images) and detected DR levels using their own customized CNN architecture and some pre-trained CNN architectures like AlexNet, and VGG-16. They classified the images into four DR stage classes. At first, the images were resized to 244 * 244 pixels and applied histogram equalization to enhance the image. Their CNN included two, two, and three CONV, max-pooling, and FC layers, respectively. Using their own CNN architecture, they got accuracy, specificity, and sensitivity of 98.15%, 97.87%, and 98.94%, respectively.

J. Wang et al. [36] worked on their private dataset (of 9194 images) and a public Messidor dataset (of 1200 images). They modified the R-FCN method [37] and detected red lesions (microaneurysms and haemorrhages) to detect DR stages in their dataset. Changing the R-FCN method is done by adding feature pyramid networks and adding five region proposal networks. In their private dataset, they got a sensitivity of 99.39%, and in the Messidor dataset, they got 92.6% sensitivity for the detection of DR stages.

Lesion Based Classification

In Lesion-based classification, DR lesions are detected, segmented and classified into certain types [44]. Adem. K et al. [38] detected exudates on public datasets DIARETDB0 (used 130 images), DIARETDB1 (used 89 images) and DrimDB (used 125 images). They used a custom CNN architecture (3 convolutional layers, three max-pooling layers, and one fully connected layer, and SoftMax is also used as a classifier) with Circular Hough Transformation (CHT). At first, they converted all datasets into grayscale images, and after that, edge detection function (Canny) and histogram equalization function (adaptive) were applied. Then, from the images, Optical discs were removed, which were detected using CHT. After that, 1152 * 1152 pixel images were fed to custom CNN as input and got accuracy, sensitivity, and specificity of 99.17%, 100%, and 98.41% on DIARETDB to detect exudates.

Y. Yan et al. [39] used DIARETDB1 [22] (89 images) and detected red lesion DR. They used the Random Forest classifier to integrate features of handcrafted and improved LeNet architecture. They used the Gaussian filter to remove the noise and used CLAHE to enhance the images. After that, U-net CNN architecture is used to segment the blood vessels from the image. Improved LeNet consists of four, three, and one CONV layer, max-pooling layer, and FC layer, respectively. In the detection of red lesions, they achieved a sensitivity of 48.7%.

J. Mo et al. [40] used the deep residual network (DRN) method of deep learning (for segmenting and classifying the exudates) on publically available datasets (E-ophtha and HEI-MED), and detected exudate lesions. A fully convolutional residual network includes down sampling and up sampling modules used in the segmentation of exudates. After that, DRN (it contains 1 CONV layer, one max-pooling layer, and five residual blocks) is used to classify exudates. To magnify the image as an input, the up-sampling module includes convolutional and de-convolutional layers. The down-sampling module consists of a CONV layer, a max-pooling layer, and 12 residual blocks. 3 CONV layers and three batch normalization layers are included in the residual block. On E-ophtha dataset, they got AUC and sensitivity of 0.9647 and 0.9227, respectively, while on HEI-MED dataset, they got AUC and sensitivity of 0.9709 and 0.9255, respectively.

Vessel Based Classification

In this method, vessels are segmented and extracted from the image, and then only DR lesions remain in the image. So, detection of remaining lesions leads to detect and classify DR images. Cam-Hao et al. [41] used the DRIVE [14] dataset and extracted retinal vessels using the pre-trained ResNet-101. Before being fed to the network, they augmented the training images first. From ResNet-101, they selected four feature maps and combined each with its neighbour. Then, they also combined feature map outputs till one feature map is obtained. After that, feature maps were concatenated at each round of best resolution. They got AUC, Accuracy, sensitivity, and specificity of 0.9732, 0.951, 0.793, and 0.9741, respectively.

Y. Wu et al. [42] worked on DRIVE [14] (used 40 images), STARE [15] (used 20 images), and CHASE [16] (used 28 images) datasets and extracted retinal blood vessels using custom CNN. At first, they converted RGB images into grayscale images, and then using CLAHE, images were enhanced and normalized. After that, for input to CNN, they extracted and augmented 48*48 patches. Their custom CNN includes 2 networks, which have skip connections and encoder-decoder structure. The encoder-decoder structure includes Convolutional layers, dropout layers, concatenation layers, and batch normalization layers. On DRIVE dataset, they achieved AUC and accuracy of 98.30% and 95.82% respectively. On STARE dataset, they got AUC and accuracy of 98.75% and 96.72% respectively. On CHASE dataset, they got AUC and accuracy of 98.94% and 96.88% respectively.

C. Tian et al. [43] used Gaussian matched filter and custom CNN architecture and extracted retinal blood vessels. They worked on DRIVE [14] (used 40 images) and CHASEDB1 [16] (used 28 images) databases. From the Gaussian filter, they acquired high and low-frequency images. After that, firstly they established a CNN path for low-frequency images (which includes convolutional layers, down-sampling, and up-sampling modules), and then they established a CNN path for high-frequency images (which includes 2 convolutional layers and 7 encoder-decoder blocks). Segmentation maps were taken out from both paths and then merged for final segmentation results. On the DRIVE dataset, they got AUC and accuracy of 0.9560 and 0.9580, respectively. On the CHASEDB1 dataset, they got AUC and accuracy of 0.9577 and 0.9601, respectively.

Conclusion

DR is mainly caused by the damage of blood vessels in the tissue of the retina, and it is one of the leading causes of visual impairment globally. To save the human eye from eyesight loss, detection of DR in early stages is necessary, and automated DR detection programs play an essential role in reducing the time, effort, and money of ophthalmologists, leading to timely treatment of DR cases. In this chapter, automated screening programs using deep learning methods for the detection and classification of DR have been discussed. Many deep learning methods are available, but most researchers have used CNN architecture only as it gives the most efficient results in medical imaging. One of the common problems of deep learning in medical imaging is the size and quality of images in datasets to train deep neural networks. This paper also reviewed the publically available datasets of retinal fundus images, but to train deep neural networks, the creation of quality retinal image databases is still in progress. Some authors have used their private datasets. In the study, we found that most of the researchers failed to detect all the stages and lesions of DR. So, an automated DR detection system that can detect all five stages and all types of lesions with high accuracy is still in need.

References

[1] Deng, L., & Yu, D. (2014). Deep learning: methods and applications. *Foundations and trends in signal processing*, 7(3–4), 197-387.

[2] Bengio, Y., Courville, A., & Vincent, P. (2013). Representation learning: A review and new perspectives. *IEEE transactions on pattern analysis and machine intelligence*, *35*(8), 1798-1828.

[3] Arel, I., Rose, D. C., & Karnowski, T. P. (2010). Deep machine learning-a new frontier in artificial intelligence research [research frontier]. *IEEE computational intelligence magazine*, *5*(4), 13-18.

[4] Schmidhuber, J. (2015). Deep learning in neural networks: An overview. *Neural networks*, *61*, 85-117.

[5] Bengio, Y., Courville, A., & Vincent, P. (2013). Representation learning: A review and new perspectives. *IEEE transactions on pattern analysis and machine intelligence*, *35*(8), 1798-1828.

[6] Arel, I., Rose, D. C., & Karnowski, T. P. (2010). Deep machine learning-a new frontier in artificial intelligence research [research frontier]. *IEEE computational intelligence magazine*, *5*(4), 13-18.

[7] Bengio, Y., Lamblin, P., Popovici, D., & Larochelle, H. (2007). Greedy layer-wise training of deep networks. *Advances in neural information processing systems*, *19*, 153.

[8] Hinton, G. E., Osindero, S., & Teh, Y. W. (2006). A fast learning algorithm for deep belief nets. *Neural computation*, *18*(7), 1527-1554.

[9] Benuwa, B. B., Zhan, Y. Z., Ghansah, B., Wornyo, D. K., & Banaseka Kataka, F. (2016). A review of deep machine learning. *International Journal of Engineering Research in Africa*, *24*, 124-136.

[10] Gupta, A., & Chhikara, R. (2018). Diabetic retinopathy: Present and past. *Procedia computer science*, *132*, 1432-1440.

[11] Alyoubi, W. L., Shalash, W. M., & Abulkhair, M. F. (2020). Diabetic retinopathy detection through deep learning techniques: A review. *Informatics in Medicine Unlocked*, 100377.

[12] Early Treatment Diabetic Retinopathy Study Research Group. (1991). Grading diabetic retinopathy from stereoscopic colour fundus photographs—an extension of the modified Airlie House classification: ETDRS report number 10. *Ophthalmology*, *98*(5), 786-806.

[13] Gadekallu, T. R., Khare, N., Bhattacharya, S., Singh, S., Reddy Maddikunta, P. K., Ra, I. H., & Alazab, M. (2020). Early detection of diabetic retinopathy using PCA-firefly based deep learning model. *Electronics*, *9*(2), 274.

[14] Niemeijer, M., Staal, J. J., Ginneken, B., Loog, M., & Abramoff, M. D. (2004). DRIVE: digital retinal images for vessel extraction. *Methods for evaluating segmentation and indexing techniques dedicated to retinal ophthalmology*.

[15] Hoover, A. D., Kouznetsova, V., & Goldbaum, M. (2000). Locating blood vessels in retinal images by piecewise threshold probing of a matched filter response. *IEEE Transactions on Medical imaging*, *19*(3), 203-210.

[16] Owen, C. G., Rudnicka, A. R., Mullen, R., Barman, S. A., Monekosso, D., Whincup, P. H., ... & Paterson, C. (2009). Measuring retinal vessel tortuosity in 10-year-old children: validation of the computer-assisted image analysis of the retina (CAIAR) program. *Investigative ophthalmology & visual science*, *50*(5), 2004-2010.

[17] Decencière, E., Zhang, X., Cazuguel, G., Lay, B., Cochener, B., Trone, C., ... & Klein, J. C. (2014). Feedback on a publicly distributed image database: the Messidor database. *Image Analysis & Stereology*, *33*(3), 231-234.

[18] Gulshan, V., Peng, L., Coram, M., Stumpe, M. C., Wu, D., Narayanaswamy, A., ... & Webster, D. R. (2016). Development and validation of a deep learning algorithm for detection of diabetic retinopathy in retinal fundus photographs. *Jama*, *316*(22), 2402-2410.

[19] *APTOS dataset* [Online]. Available, https://www.kaggle.com/c/aptos 2019-blindness-detection/data.

[20] *Kaggle dataset* [Online]. Available, https://www.kaggle.com/c/diabetic-retinopathy-detection/data.

[21] Prasanna Porwal, Samiksha Pachade, Ravi Kamble, Manesh Kokare, Girish Deshmukh, Vivek Sahasrabuddhe, Fabrice

Meriaudeau. (2018). *Indian Diabetic Retinopathy Image Dataset* (IDRiD). IEEE Dataport. https://dx.doi.org/10.21227/H25W98.

[22] Kauppi, T., Kalesnykiene, V., Kamarainen, J. K., Lensu, L., Sorri, I., Raninen, A., ... & Pietilä, J. (2007, September). The diaretdb1 diabetic retinopathy database and evaluation protocol. In *BMVC* (Vol. 1, pp. 1-10).

[23] Li, T., Gao, Y., Wang, K., Guo, S., Liu, H., & Kang, H. (2019). Diagnostic assessment of deep learning algorithms for diabetic retinopathy screening. *Information Sciences*, *501*, 511-522.

[24] Decenciere, E., Cazuguel, G., Zhang, X., Thibault, G., Klein, J. C., Meyer, F., ... & Chabouis, A. (2013). TeleOphta: Machine learning and image processing methods for teleophthalmology. *Irbm*, *34*(2), 196-203.

[25] Budai, A., Bock, R., Maier, A., Hornegger, J., & Michelson, G. (2013). Robust vessel segmentation in fundus images. *International journal of biomedical imaging*, *2013*.

[26] *ROC dataset* [Online]. Available, http://webeye.ophth.uiowa.edu/ROC/.

[27] DR2 [Online]. Available, https://figshare.com/articles/dataset/Advancing_Bag_of_Visual_Words_Representations_for_Lesion_Classification_in_Retinal_Images/953671?file=6502302.

[28] Xu, K., Feng, D., & Mi, H. (2017). Deep convolutional neural network-based early automated detection of diabetic retinopathy using fundus image. *Molecules*, *22*(12), 2054.

[29] Esfahani, M. T., Ghaderi, M., & Kafiyeh, R. (2018). Classification of diabetic and normal fundus images using new deep learning method. *Leonardo Electron J Pract Technol*, *17*(32), 233-248.

[30] Pires, R., Avila, S., Wainer, J., Valle, E., Abramoff, M. D., & Rocha, A. (2019). A data-driven approach to referable diabetic retinopathy detection. *Artificial intelligence in medicine*, *96*, 93-106.

[31] Pratt, H., Coenen, F., Broadbent, D. M., Harding, S. P., & Zheng, Y. (2016). Convolutional neural networks for diabetic retinopathy. *Procedia computer science*, *90*, 200-205.

[32] Abràmoff, M. D., Lou, Y., Erginay, A., Clarida, W., Amelon, R., Folk, J. C., & Niemeijer, M. (2016). Improved automated detection of diabetic retinopathy on a publicly available dataset through integration of deep learning. *Investigative ophthalmology & visual science*, *57*(13), 5200-5206.

[33] Gulshan, V., Peng, L., Coram, M., Stumpe, M. C., Wu, D., Narayanaswamy, A., ... & Webster, D. R. (2016). Development and validation of a deep learning algorithm for detection of diabetic retinopathy in retinal fundus photographs. *Jama*, *316*(22), 2402-2410.

[34] Shanthi, T., & Sabeenian, R. S. (2019). Modified Alexnet architecture for classification of diabetic retinopathy images. *Computers & Electrical Engineering*, *76*, 56-64.

[35] Khan, S. H., Abbas, Z., & Rizvi, S. D. (2019, February). Classification of diabetic retinopathy images based on customized CNN architecture. In *2019 Amity International Conference on Artificial Intelligence (AICAI)* (pp. 244-248). IEEE.

[36] Wang, J., Luo, J., Liu, B., Feng, R., Lu, L., & Zou, H. (2020). Automated diabetic retinopathy grading and lesion detection based on the modified R-FCN object-detection algorithm. *IET Computer Vision*, *14*(1), 1-8.

[37] Dai, J., Li, Y., He, K., & Sun, J. (2016). R-fcn: Object detection via region-based fully convolutional networks. arXiv preprint *arXiv:1605.06409*.

[38] Adem, K. (2018). Exudate detection for diabetic retinopathy with circular Hough transformation and convolutional neural networks. *Expert Systems with Applications*, *114*, 289-295.

[39] Yan, Y., Gong, J., & Liu, Y. (2019, June). A Novel Deep Learning Method for Red Lesions Detection Using Hybrid Feature. In *2019 Chinese Control and Decision Conference (CCDC)* (pp. 2287-2292). IEEE.

[40] Mo, J., Zhang, L., & Feng, Y. (2018). Exudate-based diabetic macular edema recognition in retinal images using cascaded deep residual networks. *Neurocomputing*, *290*, 161-171.

[41] Hua, C. H., Huynh-The, T., & Lee, S. (2019, July). Retinal vessel segmentation using round-wise features aggregation on bracket-shaped convolutional neural networks. In *2019 41st Annual International Conference of the IEEE Engineering in Medicine and Biology Society (EMBC)* (pp. 36-39). IEEE.

[42] Wu, Y., Xia, Y., Song, Y., Zhang, Y., & Cai, W. (2020). NFN+: A novel network followed network for retinal vessel segmentation. *Neural Networks*, *126*, 153-162.

[43] Tian, C., Fang, T., Fan, Y., & Wu, W. (2020). Multi-path convolutional neural network in fundus segmentation of blood vessels. *Biocybernetics and Biomedical Engineering*, *40*(2), 583-595.

[44] Singh, N. P., & Singh, V. P. (2020). Efficient Segmentation and Registration of Retinal Image Using Gumble Probability Distribution and BRISK Feature. *Traitement du Signal*, *37*(5).

In: Deep Learning and its Applications ISBN: 978-1-68507-185-1
Editor: Arvind Kumar Tiwari

Chapter 4

DEEP LEARNING APPROACHES FOR THE PREDICTION OF BREAST CANCER

***Sunil Dalal*[1,2,*], *Arvind Kumar Tiwari*[2] *and Jyoti Prakash Singh*[1]**

[1]Computer Science Department,
National Institute of Technology Patna, Bihar, India
[2]Computer Science Department,
Kamla Nehru Institute of Technology,
Sultanpur, U.P., India

ABSTRACT

Breast cancer is a type of cancer that develops in the cells of the breast and is a fairly prevalent disease in women. Breast cancer, like lung cancer, is a life-threatening condition for women. A promising and significant tool is automated computer technologies, particularly machine learning, to facilitate and improve medical analysis and diagnosis. Due to the great dimensionality and complexity of this data,

* Corresponding Author's E-mail: sunil.dalal@knit.ac.in.

cancer diagnosis using gene expression data remains a challenge. There is still ambiguity in the clinical diagnosis of cancer and the identification of tumor-specific markers, despite decades of study. In this study, we discuss various feature extraction techniques on different kinds of datasets. We also discuss various deep learning approaches for cancer detection and the identification of genes important for breast cancer diagnosis.

Keywords: breast cancer, deep neural network, convolutional neural network, support vector machine, feature extraction

INTRODUCTION

Cancer is a deadly disease. According to a survey, thousands of people die due to cancer every year. It is the largest cause of death in the world. It is basically a disease in which there is abnormal growth of body cells which spreads to different parts of the body. If this disease is detected in the initial stage, then this disease can be cured. Cancer basically develops due to cell growth. It originates in one part of the body and has the ability to penetrate various organs. Possible symptoms of cancer are lumps, prolonged cough, abnormal bleeding, exercise weight gain etc. Tumors are formed by most malignancies, but not all tumors are malignant. Tumors do not spread to all parts of the body. It is an abnormal growth of body tissue-when abnormal cells are stored somewhere in the body, a group of tissues is formed, which we call a tumor. These cells continue to grow abnormally and add more and more cells to their group, irrespective of the body's desire. These tumor cells are solid and fluid-filled. That process takes the form of growing cancer. This is known as metastasis. Cancer metastases are the leading cause of death-Carcinoma, melanoma, leukemia sarcoma and lymphoma are the most common cancers. Carcinomas arise in the skin, lungs, breasts, pancreas and other organs and glands. Lymphomas are lymphocyte malignancies. Leukemia [6] is a type of blood cancer. Melanomas are malignancies that develop in the cells that produce skin pigment. Breast cancer mainly occurs in women, but it is not that men cannot fall prey to it. When breast cells start to develop uncontrollably, they give rise to lumps that take the form of tumors. It depends on a genetic disease as well as many other causes. For example, if a woman consumes drugs, does not breastfeed children, children are born at an early age. For all these reasons, breast cancer is born. If a woman gets her breast examined regularly, then it can be identified at an early stage. The biggest identification of breast cancer is the presence of a lump in the breast. Mammography is becoming one of the most essential procedures for early detection of breast cancer. The most applauded alternative to a mammogram is magnetic resonance imaging (MRI). Cancer is a condition in which a cell grows abnormally. Normally, our body's cells divide and expand in a controlled manner. Normal cells die when they are injured or get old, and are replaced by healthy cells. In cancer, the signals that control cell development are malfunctioning. When cancer cells should stop growing and multiplying, they continue to do so. To put it another way, cancer cells do not adhere to the same rules as healthy cells. Jeans govern each cell, instructing it on how to function, when to grow, and when to divide. Cancer arises from the body's own cells, starting with alterations in a single cell's genes. A cell's natural function can be disrupted by genetic variations or mutations (changes or manipulations in the cell). Most translocations are harmless, although they can sometimes make it difficult to control cell development. It is uncommon for changes in genes to occur solely because the same thing happened to the parents. The majority of modifications happen on their own, as cells divide into distinct components. Not only that, but these alterations occur inadvertently and unexpectedly. Unlike cancers caused by the elderly, children's cancers are not caused by their way of life or the environment. Rather, the majority of cancer cases in youngsters are caused by genetic mutations that occur by chance. The term cancer is frequently used to describe the cell or tissue in which cancer originates. In children, there are over 100 different forms of cancer. The chemical and genetic properties of cells, as well as how cells appear under a microscope, aid in determining the type of cancer. Breast cancer has evolved into a significant disease in today's world. This condition, which brings women to the brink of death, is extremely dangerous. In India, one in every eight women is at risk of contracting the disease. According to WHO, in 2021, there will be approximately 1,78,361 breast cancer cases reported, with around 90,408 women dying as a result of breast cancer. However, if the disease is detected in time, there is a possibility of recovery from treatment.

Related Work

A support vector machine (SVM) with a dot-product kernel was utilised. Sahiner et al. [2] devised a method for extracting speculation and circumscribing margin features. Both features were quite accurate in describing bulk margins using BI-RADS descriptors. Weatherall et al. [4] proposed a method with a score of 0.93. The tumour size correlation coefficient between MRI and pathologic analysis was the best. When compared to histologic measurement, the correlation coefficients for physical exam and x-ray mammography (available for 17 patients) were 0.72 and 0.63, respectively. The MRI accuracy was unaffected by the extent of cancer residua. To see how well different imaging modalities might reliably describe the extent of a breast cancer whose location was already established. As a result, data on 20 post-chemotherapy breast cancer patients aged 32 to 66 years old was collected retrospectively. Yeung et al. [5] proposed to determine the estimations of residual tumour via each modality; the preoperative clinical and imaging records were evaluated. These results were compared to the pathologist's report's histologic measurements of the live tumour. Because of the enormous number of genes, the high quantity of noise in gene expression data, and the complexity of biological networks, it is necessary to thoroughly evaluate the raw data and utilise the relevant gene subsets. Other approaches, such as principal component analysis (PCA), have been proposed for reducing the dimensionality of expression profiles in order to help group important genes in the context of expression profiles. Bengio et al. [6] proposed Auto encoders are strong and adaptable because they extract both linear and nonlinear connections from input data. As opposed to decreasing the dimension in one step, the SDAE encoder reduces the dimensionality of the gene expression data stack by stack, resulting in less information loss. Golub et al. [8] present microarray or RNA-seq data are thoroughly explored as a classification and grouping of gene expression. Using gene expression profiles and supervised learning algorithms, numerous ways for classifying cancer cells and healthy cells have been developed. In the analysis of leukaemia cancer cells, a self-organizing map (SOM). The phases depicted in Figure 1 are followed by the majority of image processing algorithms. The screen film mammographic images must be scanned before they can be processed. One of the advantages of digital mammography is that the picture can be processed immediately. The first stage in image processing is picture pre-processing. To reduce noise and improve image quality, it must be conducted on digitised pictures. The majority of digital mammogram pictures are of high quality. If the picture is an MLO view, removing the backdrop region and the pectoral muscle from the breast area is also part of the pre-processing stage. The objective of the segmentation procedure is to discover areas of suspicious interest (ROIs), including abnormalities. In the feature extraction process, the features are computed from the attributes of the region of interest. A significant difficulty in algorithm design is the feature selection step, in which the best collection of features is chosen for preventing false positives and identifying lesion types. Choosing a smaller feature subset that delivers the highest value for a classifier performance function is referred to as feature selection. Finally, the classification stage reduces false positives and categorises lesions based on predetermined criteria.

FEATURE EXTRACTION TECHNIQUES

In the field of computer vision or image analysis, features play an important role in identifying useful information. The component picture is subjected to several picture pre-processing techniques, such as binarization, normalisation, thresholding, scaling, and so on, before picture feature extraction.

Table 1. Comparatively analysis of related work

Author	Technique used	Dataset	Performance	Feature Extraction Technique	Finding
Ragab et al. [1]	Deep CNN and SVM	DDSM with 2,256 sample and CBIS-DDSM with 5,272 sample	SVM acc 79%, Senst. 0.763, Spec 0.822, AUC 0.88, Prec 0.84, F1 score 0.8	ROI+DCNN technique	Networks will be proposed for the very deep-seated convolution network (VGG) and residual architecture (ResNet).
Sahiner et al. [2]	CNN	Radiologists obtain the Dataset from 168 mammograms	Accuracy 0.873, AUC 0.87	GLDS and SGLD technique	The CNN classifier's ability to distinguish between masses and normal ROIs obtained using an autonomous extraction approach is a first step toward a fully automated computer-aided diagnostics system.
Arbach et al. [3]	BNN	BMRI data of 20 patients	AUC 0.913 Sensitivity = 0.92 Specificity = 0.33	CAD system	Using dynamic intensity, we can improve the classification performance.
Yeung et al. [5]	k-mean + CAST + HALA	Gene expn dataset and synthetic dataset	P value 0.039 using CAST, 0.031 using k-mean and 0.049 using HALA	PCA	It worked only when PCA chose the exact right dimension, but if it failed, we can say that it purely depends on the outcome of PCA.
Bengio et al. [6]	DBNs + RBNs + supervised + unsupervised learning	UCI (split 2127 for training and 1000 for val and 1000 for test) Abalone repository dataset and second dataset is Financial	Training error 3.6%, Testing error 5.0% and validation 4.7% error	-	A simple fix network is established. We can improve by making a complex network for better improvement.
Danaee et al. [7]	ANN+SVM+ SVM-RBF	Gene expn Dataset	ANN model Acc 97.39%, Sens 96.02%, Spec 99.10%, Prec 99.17% and F-meas 0.975. SVM Model Acc 97.17%, Sens 96.38%, Spec 98.20%, Prec 98.33% and F-meas 0.973 SVM-RBF Model Acc 97.32%, Sens 89.92%, Spec 99.52%, Prec 99.58% and F-meas 0.943	SDAE features extraction technique	To assess various cancer kinds to detect biomarkers for cancer-specious c. In addition, the study of aggregated heterogeneous cancer data can reveal cross-cancer biomarkers.

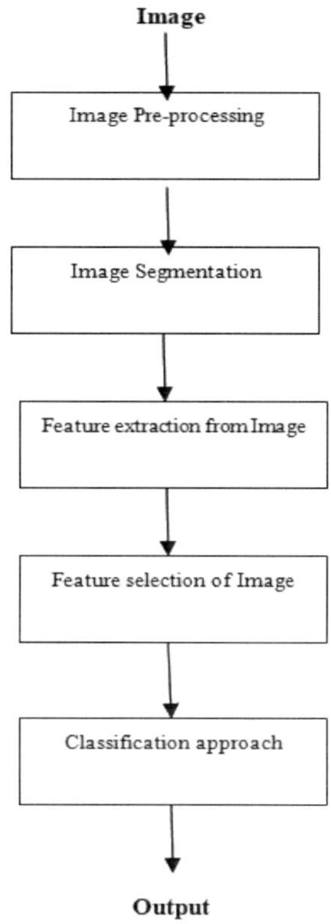

Figure 1. Typical image processing algorithm steps.

Feature extraction is the process of decreasing the amount of resources needed to explain a huge amount of data. One of the primary issues in completing complicated data analysis is the number of variables involved. GF (General features) and DSF (domain-specific features) are two types of features. FE approaches like statistical approaches can be used to extract some aspects that are not clearly recognised.

First order statistics (FOS), Gray Level Run Length Matrix (GLRLM), Gray Level Co-occurrence Matrix (GLCM), Neighbourhood Gray Tone Difference Matrix (NGTDM), and Statistical Feature Matrix are all examples of this (SFM). As illustrated in Table 2, signal processing FE approaches include law mask features, whereas transform domain approaches include Gabor wavelet, Fourier Power Spectrum (FPS) features, and discrete wavelet transform.

Deep Learning Techniques

Convolutional Neural Networks (CNNs)

Deep Learning has proven to be a particularly useful technique in recent decades due to its capacity to manage massive volumes of data. Hidden layers have eclipsed traditional approaches in popularity, particularly in pattern recognition. CNNs [11] are one of the most often used deep neural networks. The AI system, dubbed AlexNet, took second place in the 2012 ImageNet computer vision challenge with an incredible 85 percent accuracy. On the test, the runner-up received a respectable 74 percent.

Table 2. Description of selected feature extraction techniques

Various Texture / Statistical features	Features
First order statistics	Third moment, smoothness, uniformity, mean, standard deviation, entropy
GLCM features	Angular moment, inverse difference moment, contrast, correlation, entropy, difference entropy, variance, difference variance
GLRLM features	Long run emphasis, long run high gray emphasis, Low gray level run emphasis, run length non uni formity, hi gh gray level run emphasis, short run high gray emphasis, short run emphasis, short run low gray emphasis, gray level non uniformity,
GLDS features	Homogeneity, contrast, energy, entropy, mean
NGTDM	Coarsens, contrast, complexity, strength, business
SFM	Coarsens, contrast, periodicity, roughness
FPS	Radial Sum, Angular sum
Gabor filter based	Mean, variance
Shape features	Area, eccentricity, solidicity, perimeter, diameter, Euler number, orientation, convex area, extent, major axis, minor axis

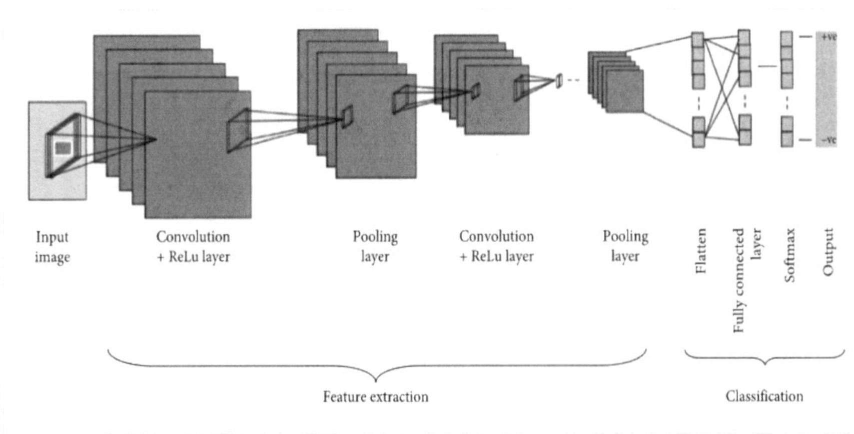

Figure 2. Typical CNN architecture for automatic detection of breast cancer.

A current CNN is usually built by layering convolutional layers on top of the input, then joining the classification output with one or many fully connected (FC) layers. To increase translational invariance and minimise feature map size, max pooling layers are frequently utilised between convolutional layers.CNNs, a form of neural network that approximates human vision, were at the heart of AlexNet. CNNs have been an integral feature of many Computer Vision applications throughout the years. As illustrated in figure 2.

Artificial Neural Networks (ANNs)

It is a computer system which has been designed to do tasks that are similar to those performed by the brain, such as creating new knowledge through learning and developing and discovering new knowledge.

It usually has three levels: an input layer, one or more hidden layers, and an output layer. Each layer has a set number of neurons or knots, which are interconnected pieces. Each neuron is linked to the others by link weights and communication linkages. On link weights, signals pass along the neurons. Each neuron receives numerous inputs proportional to their connection weights from other neurons and provides an output signal that can be generated by other neurons as well. To construct an ANN model, the network goes through two processes: training and testing. In the case of training, the network is taught to predict an output based on the input data. In the case of testing, on the other hand, the network is put to the test to see if it can stop or save training data, and it is then utilised to predict an output [12]. In addition to image processing, ultrasound, and physical assessment, a biopsy and a clear diagnosis are required to detect breast cancer. Despite the fact that a biopsy is required to provide a definite diagnosis, individuals with a low risk of cancer choose to avoid it owing to the potential for complications and expensive expenses. As a result of the aforementioned reasons, patients may choose an alternative procedure that allows them to make the right prediction before having a biopsy conducted on them. In order to achieve this aim, an ANN model was developed to assist doctors in determining if a biopsy is necessary or if the patient must be closely observed based on the patient's BI-RADS scores, age, shape of a mass, mass density, and mass boundaries.

Support Vector Machines (SVMs)

It is a supervised machine-learning tool that has a wide range of applications in classification investigations. It's been frequently utilised to solve pattern recognition, classification, and regression issues. The SVMs work on the core idea of inserting a hyper-plane between the classes and orienting it to keep it as far away from the nearest data points as possible. SVMs also have the advantage of being able to address non-linear classification issues using a kernel function. The kernel function converts two classes of data points from a lower-dimensional feature space to a higher-dimensional feature space, allowing them to be separated using a hyper-plane. To put it another way, the kernel function transforms non-linear classification issues into linear classification issues. Support Vectors are the data points that appear to be nearest to the hyper-plane [13].

Deep CNN

A positive training procedure increases the recognition rate, since data is obtained when the human body identification system has more functionality. The training procedure is based on a physiological concept that connects the interactions between neurons or properties in the training space. Each feature appears to have an essential function to perform; therefore, each feature pays attention to the scenario with its input features. During the training phase, the network employs several layers, such as the input layer, hidden layer, and output layer amongst the three layers. The input layer contains numerous sub-layer levels, notably convolution, pooling, fully connected, and normalised layers, which are used to operate attributes successfully. In the beginning, these functions were employed as a hidden layer input. The collected information is evaluated in the spread region of a convolution network. Metadata is acquired by means of a detailed learning process that enables compatibility with this source information to be recognised. The responsive field is generated by the convolution layer procedure, which results in the cluster's output and displays those very same inputs also as a cluster. The maximum gathering function is applied to the concentration layer for each cluster. This guarantees that the maximum number of features are selected from each activity cluster. The network has long been concerned about the over of data by increasing file capacity. In the pooling layer, the extracted data is integrated into the function space, and the greatest value is chosen for the fully connected layer, which computes the intrinsic merit of the output. The output importance is determined as an authorization vector in the entirely linked layer by matrix multiplication. The method iterates until the desired functions are learned and stored in the database for later recognition. To represent regular and irregular qualities, distinct individual qualities should be explored in a deep learning convolution neural network. The extracted properties are synthesised to the input level, which is then moved from the weighted input to a hidden layer to approximate the hidden layer's output [10].

CONCLUSION

One of the leading causes of mortality among women is breast cancer. Early identification and diagnosis of breast cancer can be achieved by digital mammography screening programmes, which decreases mortality and enhances the odds of complete recovery. Screening programmes generate a large number of mammographic pictures, which radiologists must interpret. Some abnormalities may be ignored or misconstrued due to the large range of appearances of breast abnormalities. Before applying any DNN technique, feature extraction techniques play an important role in identifying abnormalities and producing better results. Deep learning algorithms have the drawback of requiring big data sets that may not be accessible for cancer tissues. Researchers believe these models will increase in performance when additional gene expression data becomes accessible, revealing more interesting patterns. Deep learning models, as a result, are extremely scalable to vast input data.

REFERENCES

[1] Ragab, D. A., Sharkas, M., Marshall, S., Ren, J. Breast cancer detection using deep convolutional neural networks and support vector machines. *Peer J.,* 2019, Jan. 28; 7:e6201. doi: 10.7717/peerj.6201. PMID: 30713814; PMCID: PMC6354665.

[2] Sahiner, B., Chan, H. P., Petrick, N., Wei, D., Helvie, M. A., Adler, D. D., Goodsitt, M. M. Classification of mass and normal breast tissue: A convolution neural network classifier with spatial domain and texture images. *IEEE Trans. Med. Imaging,* 1996; 15(5):598 - 610. doi: 10.1109/42.538937. PMID: 18215941.

[3] Arbach, L. and Stolpen, A. and Reinhardt, Joseph. (2004). Classification of breast MRI lesions using a Backpropagation Neural Network (BNN). *Biomedical Imaging,* 1. 253 - 256 Vol. 1. 10.1109/ ISBI.2004.1398522.

[4] Weatherall, P. T., Evans, G. F., Metzger, G. J., Saborrian, M. H., Leitch, A. M. MRI vs. histologic measurement of breast cancer following chemotherapy: Comparison with x-ray mammography and palpation. *J. Magn. Reson Imaging,* 2001, Jun.; 13(6):868 - 75. doi: 10.1002/jmri.1124. PMID: 11382946.

[5] Yeung, Ka Yee and Ruzzo, Walter. (2001). Principal Component Analysis for clustering gene expression data. *Bioinformatics,* (Oxford, England). 17. 763 - 74. 10.1093/bioinformatics/17.9.763.

[6] Bengio, Y. and Lamblin, Pascal and Popovici, Dan and Larochelle, Hugo and Montreal, U. (2007). Greedy layer-wise training of deep networks. *Advances in Neural Information Processing Systems,* 19.

[7] Danaee, P., Ghaeini, R., Hendrix, Da. A Deep Learning Approach for Cancer Detection and Relevant Gene Identification. *Pac. Symp. Biocomput.*, 2017; 22:219 - 229. doi: 10.1142/9789813207813_ 0022. PMID: 27896977; PMCID: PMC5177447.

[8] Golub, T. R., Slonim, D. K., Tamayo, P., Huard, C., Gaasenbeek, M., Mesirov, J. P., Coller, H., Loh, M. L., Downing, J. R., Caligiuri, M. A., Bloomfield, C. D., Lander, E. S. Molecular classification of cancer: Class discovery and class prediction by gene expression monitoring. *Science,* 1999, Oct. 15; 286(5439):531 - 7. doi: 10.1126/ science.286.5439.531. PMID: 10521349.

[9] Balaji Krishnapuram, Lawrence Carin and Hartemink, Alexander J. 2003. Joint classifier and feature optimization for cancer diagnosis using gene expression data. In <i>*Proceedings of the seventh annual international conference on Research in computational molecular biology.* Association for Computing Machinery, New York, NY, USA, 167 - 175. DOI:https://doi.org/10.1145/ 640075.640097.

[10] Ratnala Venkata Siva Harish, Dr. P. Rajesh Kumar. 2020. A Deep Convolutional Neural Network (DCNN) and Squirrel Search Algorithm (SSA) based Classifier Framework by Extracting Human Body Skeleton points based on Silhouette Images for Human Action Recognition in Image Processing. *International Journal of Future Generation Communication and Networking,* Vol. 13 No. 3 (2020).

[11] Alanazi, S. A., Kamruzzaman, M. M., Islam Sarker, M. N., Alruwaili, M., Alhwaiti, Y., Alshammari, N., Siddiqi, M. H. Boosting Breast Cancer Detection Using Convolutional Neural Network. *J. Healthc. Eng.,* 2021, Apr. 3; 2021:5528622. doi: 10.1155/2021/5528622. PMID: 33884157; PMCID: PMC8041556.

[12] Saritas, I. Prediction of breast cancer using artificial neural networks. *J. Med. Syst.,* 2012, Oct; 36(5):2901 - 7. doi:

10.1007/s10916-011-9768-0. Epub 2011 Aug. 12. PMID: 21837454.

[13] Sewak, M., Vaidya, P., Chan, C. and Zhong-Hui Duan. "SVM Approach to Breast Cancer Classification", *Second International Multi-Symposiums on Computer and Computational Sciences (IMSCCS 2007),* 2007, pp. 32 - 37, doi: 10.1109/IMSCCS.2007.46.

In: Deep Learning and its Applications
Editor: Arvind Kumar Tiwari
ISBN: 978-1-68507-185-1

Chapter 5

DEEP LEARNING TECHNIQUES FOR THE PREDICTION OF EPILEPSY

Anuj Singh* and Arvind Kumar Tiwari†
Department of Computer Science and Engineering,
Kamla Nehru Institute of Technology, Sultanpur, India

ABSTRACT

Deep learning helps simulation techniques with various computing layers to gain several stages of abstraction for data representations. These techniques have vastly enhanced the position in voice detection, visual target recognition, particle identification as well as a variety of other fields including drug discovery as well as genomics. Deep learning uses the backpropagation method to show how a computer can adjust the input variables that are employed to measure the value in every layer from the description in the subsequent layer revealing detailed structure in huge volumes of data. Deep learning has accomplished substantial progress as well as demonstrated outstanding effectiveness in a variety of applications like Adaptive testing, cancer detection, natural language processing, face recognition, speech

* Corresponding Author's Email: anujsinghknit@gmail.com.
† Corresponding Author's Email: arvind.tiwari1980@gmail.com.

recognition, and much more. Epileptic seizures are chronic neurological disorders with serious public health consequences. Epileptic seizures are a type of neurological disorder that can affect children between the ages of 10 to 20, as well as adults between the ages of 65 to 70. Pre-ictal, Ictal, Post-ictal, and Interictal epileptic seizures are the four phases that can be analyzed. The basic goal of this article is to offer a sequential overview of the primary uses of deep learning in a range of fields, including a study of the methodologies and architectures employed as well as the impact of each implementation in the actual world. This chapter shows that ANN had 98.26% accuracy, CNN had 97.52% accuracy, and LSTM had 95.78% accuracy. In terms of F1-Score, we found that ANN has a score of 95.57%, CNN has a score of 93.77%, and LSTM has a score of 89.87%. In addition, when compared to previous techniques, the convolutional neural networks model performed better.

Keywords: deep learning, convolutional neural network, recurrent neural network, long short term memory, generative adversarial network, epileptic seizures.

INTRODUCTION

Deep learning helps simulation techniques with various computing layers to gain several stages of abstraction for data representations. These techniques have vastly enhanced the position in voice detection, visual target recognition, particle identification as well as a variety of other fields including drug discovery as well as genomics. Deep learning uses the backpropagation method to show how a computer can adjust the input variables that are employed to measure the value in every layer from the description in the subsequent layer revealing detailed structure in huge volumes of data [1]. Deep learning is perhaps the highest accuracy, supervised as well as time and cost-effective machine learning method. Deep learning is not a limited learning methodology rather it encompasses a wide range of methods that can be employed in a wide range of complex situations [2].

Artificial Intelligence

Let's start with a definition of intelligence. Intelligence is defined as the ability to learn and solve issues. Its primary goal is to create computers so clever that they can act intelligently in the same way that humans do. If a computer learns new information, it can intelligently solve real-world problems based on previous experiences. As we all know, intelligence is the capacity to learn and solve issues, and intelligence is attained via knowledge, which is attained in part via information, which is attained via prior experiences and experiences obtained through training. Finally, by combining all of the elements, we can conclude that artificial intelligence can obtain knowledge and apply it to execute tasks intelligently based on their previous experiences. Reasoning, learning, problem solving, and perception are all aspects of intelligence. Artificial intelligence systems are required to minimize human workload. Natural Language Processing (NLP), Speech Recognition, Healthcare, Vision Systems, and Automotive are just a few of the applications. An agent and its surroundings make up an Artificial Intelligence system. The environment is perceived by sensors, and the environment is reacted to by effectors. An intelligent agent sets objectives and is extremely interested in achieving them. Artificial Intelligence has developed some tools to help handle tough and complicated issues, including neural networks, languages, search and optimization, and uncertain reasoning, among others.

Machine Learning

Machine learning is the subset of artificial intelligence (AI). Machine learning, as the name implies, is the capacity of a machine to learn. Machine learning is a branch of computer science that allows computers to learn and solve problems without being explicitly programmed. Here, we create a machine that performs duties similar to those performed by humans to decrease human labor. It is a branch of research that enables computers to learn new things using information fed to them and to produce more efficient output using that information. It is employed in a variety of professions and has gained notoriety in a variety of sectors. It is a fantastic technology that allows machines or computers to learn and solve complicated problems.

DEEP LEARNING

Deep learning is a subset of artificial intelligence and machine learning. It is based entirely on Artificial Neural Networks. We already knew that neural networks can artificially mimic the human brain. As a result, deep learning copies the neural network to artificially mimic the human brain. We don't use explicit programming here. The concept of deep learning is rather ancient, but it is quite popular these days since processing power was not very high previously, and we also didn't have nearly as much data. However, computing power has increased at an exponential rate in the previous two decades. As a result, machine learning and deep learning are becoming increasingly popular. There are roughly 100 billion neurons in the human brain that are linked. Each neuron is linked to about 1000 neurons in its immediate vicinity. The major challenge now is how we can link these neurons in a computer. The invention of Artificial Neural Networks is now the most well-known answer to this challenge. There are three types of layers in an artificial neural network: input layer, an output layer, and hidden layers in between these two layers. Deep Belief Network, Deep Neural Network, Convolutional Neural Network, and Recurrent Neural Network are examples of deep learning architectures. Image identification, speech recognition, audio recognition, medical image analysis, machine recognition, and natural language processing (NLP) are just a few of the applications for deep learning systems. Deep learning architectures do well in several disciplines when compared to human performance. The human brain was heavily influenced by artificial neural networks. The artificial neural network and the organic human brain have several distinctions. Artificial neural networks are non-living, static, and provide symbolic representations of multiple neurons, whereas the biological human brain is a living creature that is dynamic and analog. The term "deep" in deep learning refers to the fact that the network has several hidden layers in addition to the input and output layers. The network will learn more and provide more exact results as the number of hidden layers grows. The number of layers in deep learning is infinite, but the size of each layer is constrained. Deep learning is a subset of machine learning that uses numerous hidden layers to extract characteristics from raw input. As the number of hidden layers grows, the model produces results that are increasingly exact and accurate, similar to those of an expert person. If we use a deep neural network model in an image processing system, the first hidden layer just extracts edges from the raw input photos, but the last hidden layer produces clear and accurate results at the level of human comprehensions, such as faces, characters, or numerals.

Deep Learning Models

The term "deep learning" comes first from artificial neural networks [3]. A convolutional neural network is perhaps the most essential of the several deep learning networks since it considerably encourages the growth of image analysis. Generative adversarial network as a novel deep learning model offers up new boundaries for the study as well as application of deep learning which has lately received a lot of attention [4].

Convolutional Neural Network

The neural network image processing group was the first to create the convolutional neural network. Convolutional neural networks are a type of deep neural network used in deep learning to analyze pictures.

Convolutional neural networks are extensively employed in image identification, object identification, picture classifications, and face identification, among other applications. Convolution takes an input picture, works on it, and then uses the values to categorize it (either cat or dog, pen or pencil). ConvNets are developed to accommodate data in the form of several arrays, such as a color image made up of three 2D arrays representing pixel elevations in each of the color channels. As attribute extractors, a CNN uses two operations as convolution and pooling. As in a multi-layer perceptron, the output of this series of operations is bound to a completely connected layer. Convolutional neural networks are often used on text in Natural Language Processing. There are two types of pooling used: max-pooling and average-pooling. When we use CNN for text instead of images, we display the text with a 1-Dimensional string. CNN is mostly used in sentence classification in NLP tasks. Microsoft earlier released a range of optical character recognition software, as well as handwriting recognition software, specifically focuses on ConvNet [5]. ConvNets were still used to recognize objects in real photographs, such as faces as well as legs, and even to recognize faces in the early 1990s [6].

Recurrent Neural Network

The biggest thrilling utilization of backpropagation since it was initially developed was for recurrent neural networks (RNNs) simulation. RNNs are also preferred for functions that need sequential inputs, like expression and vocabulary. It produces the output based on previous computation by using sequential information. Recurrent Neural Networks are similar to neural networks, however, they do not function in the same way. As humans, we do not think from the ground up. For example, if we're watching a movie, we may guess what will happen next based on what we know about the prior one. A typical neural network, on the other hand, is unable to predict the next action in the film. Recurrent Neural Networks can address challenges like these. In a recurrent neural network, there is a loop in the network that keeps the data. It can take more than one input vector and produce more than one output vector. Recurrent neural networks use memory cells that are capable to capture information about long sequences. RNNs are extremely strong dynamic structures, although teaching them has proven difficult because backpropagated gradients expand or retreat at every time stage, causing them to burst or disappear over several time stages [7].

LONG SHORT TERM MEMORY

Sepp Hochreiter and Juergen Schmidhuber presented another model for sequential information called long short-term memory. LSTM networks are a form of RNN that may learn order dependence in situations of sequence prediction. The LSTM architecture is an RNN that remembers information at regular periods. It is employed in the solution of the vanishing gradient problem. It can learn long-term dependence. At observation time, RNN has only two gates: an input gate, as well as an output gate from the last hidden state and there, is no knowledge about the past to remember. RNNs can remember their inputs for a long time due to LSTMs. That is why long short-term memory uses its memory to accumulate information over a long time. This memory cell is known as a gated cell because it represents whether or not to store or delete information dependent on the relevance of the information. LSTM is made up of three gates. The input gate is used for new data input, the forget gate is used to determine whether or not the data should be deleted, and the output gate is used to determine the output at the current time step. Some of the applications of LSTM are robot control [8], Time series prediction [9], rhythm learning [10], human action recognition [11], semantic parsing [12], sign language translation [13], drug design [14], market prediction [15], and protein homology detection [16].

Generative Adversarial Network

Generative adversarial network (GAN) is a novel deep learning concept that includes a unique neural network model that trains generator as well as discriminator at the same time. The generator's job is to know and understand the probability distribution of actual images and afterward add random noise to them to make false images, whereas the discriminator's job is to determine whether a source variable is genuine or not [17]. The discriminator, as well as generator, has been tweaked to enhance their performance. The generative adversarial network training process is unique in that it has been using backpropagation to train as well as utilizes the confrontation of two neural networks as training metric, significantly reducing the training problem as well as improving the training effectiveness of the induced model. A generative adversarial network provides an opportunity to learn deep representation without having to label your training data extensively [18]. One of the most often used generative adversarial network applications is computer vision [19].

Epileptic Seizures

Epilepsy is a neurological disease that is persistent, severe, and debilitating. About 1-2% of the world's population has epilepsy, with nearly 30-40% of those suffering from medication-resistant epilepsy. Epileptic seizures can be anticipated, according to research, since individuals experience "Auras" before they have an epileptic seizure. Inter-ictal, Pre-ictal, Ictal, and Post-ictal are the four phases of epileptic seizures. The Inter-ictal stage is the period between epileptic seizures, whereas the Pre-ictal stage is the period just before the epileptic seizure, the Ictal stage is the epileptic episode itself, and the Post-ictal stage is the period immediately after the epileptic seizure. Since there are four seizure states, we are only interested in two of them: inter-ictal and pre-ictal. Because they are not relevant to our topic, ictal and post-ictal states are excluded. Now we'll look at how to distinguish between inter-ictal and pre-ictal states. As a result, the work of seizure prediction becomes a classification problem [20]. Drugs or surgical techniques can be used to treat epileptic individuals. The epileptic patient benefits from earlier predictions in that they can protect themselves before suffering serious damage or attack by taking the proper medication and improving their condition[21]. Only half of the epileptic patients who have surgery can avoid having an epileptic seizure. The rest individuals who have unpredictably unexpected seizures throughout their daily schedule may have a very severe episode. It has been noted that total therapy of epileptic seizures is not possible. Many scientists are attempting to develop procedures that will allow seizures to be anticipated earlier they occur [22]. Because capturing brain electric activity is challenging, numerous researchers have turned to ECG signals. The rationale for this is that medical evidence shows that the heart rate pattern changes before an epileptic seizure [23].

An electroencephalogram (EEG) is a test that records the electrical activity of the brain (signals). An EEG was initially used by a German doctor to record the electrical activity of the human brain [24]. Even the finest seizure prediction system hasn't been able to provide good sensitivity (the ability to predict seizure) and specificity (the ability to ignore false alarms) in the last 30 years [25]. Annually, over 2 million fresh epileptic patient accounts are created, with 70 percent of epileptic patients being able to restore their health using anti-epileptic drugs (AEDs) and the other 30 percent being uncontrollable. Seizures can cause loss of recognition or consciousness, difficulty moving, impaired perception, and loss of other thinking abilities. Due to various uncontrolled seizures, epileptic patients are prohibited from riding or driving any vehicles and are unable to get a genuine or good job. Children between the ages of 10 to 20 years old, as well as individuals between the ages of 65 to 70 years old, have been reported to suffer from epilepsy. The most successful and powerful instrument for capturing brain electric activity is the electroencephalogram (EEG). It is extremely difficult to capture electrical signals from the brain without the use of an

EEG. True worry is aroused if the pre-ictal state is discovered among the primary inter-ictal states [26]. To record EEG, electrodes are inserted on the patient's scalp. The brain's electrical activity may be recorded in this way. The brain wave samples differ between the pre-ictal and inter-ictal stages [27]. As a result, bivariate characteristics are used instead of univariate data in seizure prediction techniques. Because there are variations in seizure kind and location across patients, seizure prediction systems are customized for each patient. As a result, the EEG signals of the patients differ. The authors utilized a supervised machine learning methodology that focuses on two critical stages: feature extraction and categorization between pre-ictal and inter-ictal phases. On the retrieved feature of the EEG data, a multilayer perceptron and a convolutional neural network were utilized to anticipate epileptic seizures. Because it's critical to predicting epileptic seizures ahead of time and with precision, the scientists created a novel automated system that combines feature extraction and classification. The authors created a deep learning-based system that extracts significant characteristics without the need for any pre-processing. To extract features, a multilayer perceptron and a deep convolutional neural network were utilized. To do classification, a deep convolutional neural network is paired with a recurrent neural network.

ELECTROENCEPHALOGRAM (EEG)

The electrical activity of the brain is recorded using EEG. It is a method of electrophysiological monitoring. When electrodes are positioned along the scalp area, it becomes non-invasive, and when electrodes are utilized like in electrocorticography it becomes invasive. It detects voltage changes caused by an ionic current within a neuron in the brain. It refers to the use of many electrodes on the scalp to capture the brain's unrestrained electrical activity regularly. EEG recording is primarily utilized for improved epilepsy therapy [28]. The most helpful instrument for measuring brain activity is the electroencephalogram. The anomalies in the brain detected by electroencephalogram were caused by epilepsy. EEG reading refers to the recording of brain function; any irregularities may be quickly seen. In the treatment of strokes, brain diseases, and cancers, EEG is utilized as the first or most significant approach. However, with the development of high-resolution imaging techniques such as MRI (Magnetic Resonance Imaging) and CT (Computerized Tomography), its popularity has waned (Computed Tomography). Despite its limited spatial resolution, EEG has always been a valuable or worthwhile tool for diagnosis and study. It has a temporal resolution of milliseconds, which Magnetic Resonance Imaging, Positron Emission Tomography, and Computed Tomography can't match [29].

APPLICATION OF ELECTROENCEPHALOGRAM (EEG)

The EEG is the most significant, valuable, or priceless tool for an epilepsy diagnosis. It takes roughly 20-30 minutes to read the patient's brain regularly. The brain of an epileptic patient must be clinically recorded regularly. It takes about 20-30 minutes to record an EEG. It lasts between 20 and 30 minutes. The electrodes are placed on the scalp to monitor the electrical activity of the brain. Small metal discs serve as electrodes. If any anomalies are discovered in the EEG recording, a diagnosis of brain illnesses is required. Patients with epilepsy should have routine testing done regularly. If epileptic patients' routine checkups are not done properly, we will witness a significant loss, which may result in the epileptic patient's death. If the patient's epilepsy is not properly diagnosed, it can be exceedingly severe or destructive [30]. EEG is crucial for diagnosing a variety of brain illnesses, including tumors, brain damage from a head injury, sleep problems, encephalitis, encephalopathy, and stroke. Recent research using machine learning approaches such as neural networks and mathematics such as statistical temporal characteristics taken from frontal lobe EEG brainwave recordings has shown notable results in categorizing mental conditions [31].

Epilepsy Symptoms

It depends on the sort of epileptic seizure you're having. Because the brain activity during epilepsy is abnormal. Because it directly affects the patient's neurological system, brain activity becomes aberrant. As a result, epileptic seizures can impair any process that is coordinated by the brain. Confusion, fear, anxiety, loss of consciousness, uncontrollable jerking movements, and loss of awareness are some of the usual symptoms. Genetic Influence, Head Trauma, Development Disorders, Prenatal Injury, Brain Conditions, and Infectious Diseases, among other reasons, are unknown to half of those who suffer from epilepsy [32].

Related Work

In paper [33], the author developed a new method of identifying current IR models, as well as their massive advances and innovations. This method was the first one to categorize established tasks based on how the features, as well as ranking functions, are formed. In paper [34], the author conducted the first complete investigation on the implementation flaws of mobile deep learning applications. They found 304 genuine implementation flaws using datasets from Stack Overflow as well as GitHub popular datasets for analyzing software flaws. They created a perfect taxonomy based on the discovered defects, which includes 23 classes for malfunction symptoms as well as distills common remedy procedures for various fault kinds. The authors of the article [35], created a hybrid deep-learning model focused on the convolutional neural networks as well as gcForest. The CWT was used to generate bearing vibration data into time-frequency graphics. The images were then fed into a gcForest classifier using a convolutional neural network to identify inherent defect characteristics. In paper [36], the authors provided an outline of several most often employed deep learning architectures, relevant libraries, as well as several practical domains in which deep learning is applied. In paper [37], the authors provided an overview of the origin, evolution, as well as uses of deep learning methods, with a focus on their use in medical imaging. In the paper [38], the authors discussed the deep learning research for medical imaging synthesis including their practical diagnosis. They listed or highlighted the suggested approaches, research designs, as well as observed outcomes with relevant medical implications on sample investigations to describe the latest advancements of deep learning-based approaches in intermodality as well as intra modality picture synthesis. The author of the article [39], performed a careful assessment of current deep learning models' applications on machine healthcare monitoring activities as well as provided their observations on these models. On a difficult platform wear estimation problem, realistic research of traditional machine learning models, as well as deep learning models, have been presented. The author of the article [40], provided a complete overview of frequently employed deep learning models as well as discussed how they may be utilized to make manufacturing "smart." First, the evolution of deep learning methods is described, as well as their merits above classical machine learning. Following that, deep learning-based cognitive algorithms are provided to improve performance in manufacturing. The authors of article [41], evaluated different approaches to offer clinical reasoning capable of justifying the effectiveness of deep learning as well as they believe that this is the best viable path to go, even though present successes are too fragmented and useful for just a few classes of deep neural networks. According to the authors, the majority of what can describe why deep learning performs at all much alone extremely effectively, throughout several fields of application is still unknown, as well as a further study into the theoretical foundations of artificial learning is urgently desired.

In paper [42], the authors illustrated the usefulness of a modified form of deep ResNet for examining neuroimaging signals in the target case of forecasting evolution from moderate cognitive impairment to alzheimer's disease. The deep models were initially trained on moderate cognitive impairment persons solely, then by a domain transfer learning variant which subsequently trained on alzheimer's disease and controls for forecasting. They furthermore show how to locate irregularities using a network occlusion approach. The methods investigated demonstrated deep learning architecture's strong potential for learning subtle predictive characteristics as well as use in essential areas like illness advancement prediction or comprehension. The authors in article [43], used deep learning to construct imaging characteristics from pretreatment diffusion-weighted image datasets as well as tested their capacity to forecast diagnostic results in individuals with major artery blockage. In paper [44], the authors presented a synergic deep learning approach to solve the problems of a variety of imaging modalities and clinical diseases by deploying several DCNNs at the same time along with allowing them to train from one another. The learned image representations of every set of deep convolutional neural networks (DCNNs) are combined as the source to a synergic network, that has a completely integrated architecture that forecasts if the pair of source images corresponds to a single class. This method can be trained from start to finish using deep convolutional neural network classification problems as well as synergic mistakes from every pair of deep convolutional neural networks. The authors of the article [45], presented a method for predicting seizures based on the individual patient. Prediction of epileptic seizures for a certain patient is done here. The intracranial electroencephalogram signal was employed in this approach to predict epileptic seizures in people and dogs. The best result for human seizure prediction was Time window Tw=60 sec (AUC=0.9349); the best result for dog seizure prediction was Time window Tw=30 sec (AUC=0.9432). The author of the article[46], evaluated the level of loss of responsiveness in those who have psychogenic non-epileptic seizures, as well as the physical changes that might be linked to this type of seizure. Loss of reactivity is common in patients with psychogenic non-epileptic seizures, and it is linked to seizure-related damage. The authors of the paper [47], demonstrate how applying functional Near-Infrared Spectroscopy (fNIRS) to the identification of epileptic seizures yields results that are superior to those based on EEG and that the deep learning method to this problem is effective when functional Near-Infrared Spectroscopy recordings are available.

Feature Selection

To reduce the dataset's exceptionally high dimensionality, feature selection was critical. Improve prediction performance, create faster and more efficient models, and provide a deeper understanding of the underlying processes are some of the most important goals of feature selection. If our dataset has a large number of features, this does not automatically imply that it will provide relevant info or insights. Since several characteristics are useless, a high amount of features may not always provide relevant info [48]. The term "redundant" refers to the fact that the same features appear again in our dataset. This might result in memory and processing power is wasted. Overfitting can occur when a high number of redundant characteristics are used in machine learning. It has been discovered that feature extraction is the superior technique for resolving this issue. Feature extraction's major purpose is to minimize the number of features in an existing dataset and utilize those minimized features in the model rather than the entire original dataset [49].

Methodology

Traditional machine learning algorithms cannot extract better feature representation from unstructured data than CNN. It has also been shown to be effective. Because EEG data is one-dimensional time-series data, a one-dimensional convolution neural network works well in such type of data. The status of the result in tasks such as Image Classification, Speech Recognition, and many Natural Language problems has been provided by convolution neural networks. Traditional machine learning algorithms cannot extract better feature representation from unstructured data than convolution neural networks. It has also been shown to be a reliable way of extracting features. The convolution neural network allows input neurons to communicate sparsely with output neurons. To do this, CNN employs a convolutional algorithm. Sparse interaction is usually achieved by using a kernel that is smaller than the feature size. In comparison to a fully linked network, sparse interaction allows us to keep fewer hyperparameters, which saves memory and time. The dataset utilized in this chapter is from the kaggle website's epileptic seizure prediction competition. On the kaggle website, this dataset is freely accessible. The core dataset is divided into five folders, each of which contains 100 files, each representing a single subject per person. And it should be emphasized that the brain activity is recorded for 23.6 seconds, and this information is incorporated in each file. The parallel time series is sampled with 4097 data points. Each piece of information reflects the EEG recording value of a different participant at a different moment. We have 500 separate files, which implies 500 distinct people, each with 4097 data points reflecting the value of 23.5 seconds of EEG recording. Now we'll divide and reorganize the 4097 data points into 23 pieces, each with 500 files. Each chunk has 178 data points, each indicating the value of EEG recording at a different moment. So we have 23*500=11500 rows, each of which has 178 data points for 1 second, corresponding to 178 columns, with the last column entitled 'class' having binary values of Yes or No. Yes indicates that the person has had an epileptic seizure, whereas No indicates that the person has not had an epileptic seizure.

PERFORMANCE EVALUATION

Specificity, sensitivity, accuracy, precision, and f-measure are some of the measures used to evaluate performance. True negative (TN), false negative (FN), true positive (TP), and false-positive (FP) are used to calculate performance (FP).

TP = Number of true values that are correctly identified.
FP = Number of false values that are identified as True
TN = Number of false values that are correctly identified
FN = Number of true values that are incorrectly identified

Confusion Matrix

It is a very basic matrix that is used to determine the model's accuracy and other critical metrics. It is mostly used to solve classification problems in which the output may be divided into two or more classes [50].

Table 1. Confusion matrix

	Negative (Predicted)	Positive (Predicted)
Negative (Actual)	True Negative(TN)	False Positive(FP)
Positive (Actual)	False Negative(FN)	True Positive(TP)

Evaluation Parameters

We provide a set of assessment measures for prediction accuracy that is the most widely employed. The specificity, accuracy, recall, precision, and F-score metrics are used to assess the tests' performance. These five classic measures are also used to assess a wide range of diagnostic tests.

Accuracy

Correct predictions as True Positive and True Negative in the numerator, and all of the model's predictions in the denominator. It's crucial in the classification problem, and it's defined as the ratio of the model's right predictions to the total number of predictions produced [51].

$$\text{Accuracy} = \frac{\text{true positive+true negative}}{\text{true positive+true negative+false positive+false negative}} \quad (1)$$

Precision

The percent of relevant instances among all retrieved instances is known as precision or positive predictive value. The relevance of the result set is tested using precision [52].

$$\text{Precision} = \frac{\text{true positive}}{\text{true positive+false positive}} \quad (2)$$

Recall

The recall is the percentage of relevant documents that will be successfully retrieved from all the documents that have been retrieved [53].

$$\text{Recall} = \frac{\text{true positive}}{\text{true positive+false negative}} \quad (3)$$

F-Measure

To calculate the result, the F-measure considers the test's precision and recall. The precision, also known as a positive predictive value, is the percentage of relevant instances found among all retrieved instances, and recall is the percentage of relevant documents found among all retrieved documents that are successfully retrieved. The harmonic mean of precision and recall is the F-measure [54].

$$F = \frac{2*precision*recall}{precision+recall} \quad (4)$$

Specificity

It is exactly the opposite of the Recall. It is a measure that describes the proportion of True Negative concerning all negative data points [55].

$$\text{Specificity} = \frac{true\ negative}{true\ negative+false\ positive} \quad (5)$$

Result Analysis

We use the Epileptic Seizure dataset in this work to determine the various performance measurement parameters of several deep learning techniques. On the presented dataset of epileptic seizures, we used multiple methods to determine various parameters. We compared the Artificial Neural Network (ANN), Convolutional Neural Network (CNN), and Long Short Term Memory (LSTM) deep learning models. We found that the Convolutional Neural Network (CNN) produces the best results in this work (See Table 2).

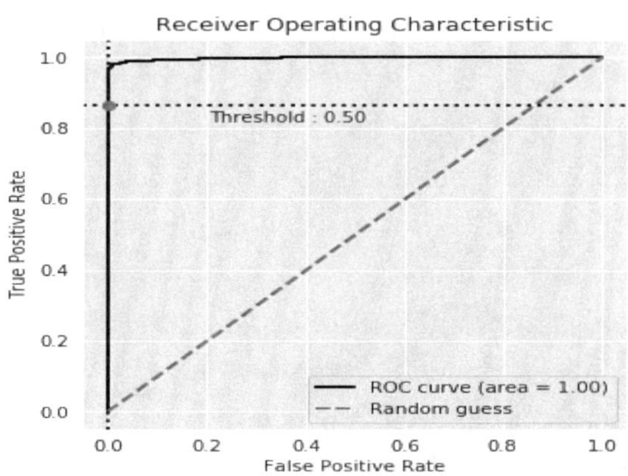

Figure 1. Receiver operating characteristic curve (ROC) for the classification using Convolutional Neural Network (CNN).

Table 2. The evaluation metrics for epileptic dataset

Algorithms	ANN	CNN	LSTM
Accuracy (%)	94.96%	97.22%	95.78%
Precision (%)	98.91%	99.76%	89.40%
Recall (%)	76.47%	86.76%	90.34%
F1-Score (%)	86.25%	92.81%	89.87%
Specificity (%)	99.78%	99.95%	97.20%

We found that ANN had 98.26% accuracy, CNN had 97.52% accuracy, and LSTM had 95.78% accuracy. In terms of F1-Score, we found that ANN has a score of 95.57%, CNN has a score of 93.77%, and LSTM has a score of 89.87%.

CONCLUSION

Deep learning helps simulation techniques with various computing layers to gain several stages of abstraction for data representations. These techniques have vastly enhanced the position in voice detection, visual target recognition, particle identification as well as a variety of other fields including drug discovery as well as genomics. Deep learning uses the backpropagation method to show how a computer can adjust the input variables that are employed to measure the value in every layer from the description in the subsequent layer revealing detailed structure in huge volumes of data. Deep learning has accomplished substantial progress as well as demonstrated outstanding effectiveness in a variety of applications like Adaptive testing, cancer detection, natural language processing, face recognition, speech recognition, and much more. Epileptic seizures are chronic neurological disorders with serious public health consequences. Epileptic seizures are a type of neurological disorder that can affect children between the ages of 10 to 20, as well as adults between the ages of 65 to 70. Pre-ictal, Ictal, Post-ictal, and Interictal epileptic seizures are the four phases that can be analyzed. The basic goal of this article is to offer a sequential overview of the primary uses of deep learning in a range of fields, including a study of the methodologies and architectures employed as well as the impact of each implementation in the actual world. In the proposed work we found that ANN had 98.26% accuracy, CNN had 97.52% accuracy, and LSTM had 95.78% accuracy. In terms of F1-Score, we found that ANN has a score of 95.57%, CNN has a score of 93.77%, and LSTM has a score of 89.87%. In addition, when compared to previous techniques, the convolutional neural networks model performed better.

REFERENCES

[1] LeCun, Yann, Yoshua Bengio, and Geoffrey Hinton. “Deep learning.” *Nature* 521, no. 7553 (2015): 436-444.

[2] Dargan, Shaveta, Munish Kumar, Maruthi Rohit Ayyagari, and Gulshan Kumar. “A survey of deep learning and its applications: A new paradigm to machine learning.” *Archives of Computational Methods in Engineering* (2019): 1-22.

[3] Hopfield, John J. “Neural networks and physical systems with emergent collective computational abilities.” *Proceedings of the national academy of sciences* 79, no. 8 (1982): 2554-2558.

[4] Ledig, Christian, Lucas Theis, Ferenc Huszár, Jose Caballero, Andrew Cunningham, Alejandro Acosta, Andrew Aitken et al. “Photo-realistic single image super-resolution using a generative adversarial network.” In *Proceedings of the IEEE conference on computer vision and pattern recognition,* pp. 4681-4690. 2017.

[5] Simard, Patrice Y., David Steinkraus, and John C. Platt. “Best practices for convolutional neural networks applied to visual document analysis.” In *ICDAR,* vol. 3, no. 2003. 2003.

[6] Lawrence, Steve, C. Lee Giles, Ah Chung Tsoi, and Andrew D. Back. “Face recognition: A convolutional neural-network approach.” *IEEE transactions on neural networks* 8, no. 1 (1997): 98-113.

[7] Bengio, Yoshua, Patrice Simard, and Paolo Frasconi. "Learning long-term dependencies with gradient descent is difficult." *IEEE transactions on neural networks* 5, no. 2 (1994): 157-166.

[8] Mayer, Hermann, Faustino Gomez, Daan Wierstra, Istvan Nagy, Alois Knoll, and Jürgen Schmidhuber. "A system for robotic heart surgery that learns to tie knots using recurrent neural networks." *Advanced Robotics* 22, no. 13-14 (2008): 1521-1537.

[9] Schmidhuber, Jürgen, Daan Wierstra, and Faustino J. Gomez. "Evolino: Hybrid neuroevolution/optimal linear search for sequence prediction." In *Proceedings of the 19th International Joint Conference on Artificial Intelligence (IJCAI).* 2005.

[10] Gers, Felix A., Nicol N. Schraudolph, and Jürgen Schmidhuber. "Learning precise timing with LSTM recurrent networks." *Journal of machine learning research* 3, no. Aug (2002): 115-143.

[11] Baccouche, Moez, Franck Mamalet, Christian Wolf, Christophe Garcia, and Atilla Baskurt. "Sequential deep learning for human action recognition." In *International workshop on human behavior understanding,* pp. 29-39. Springer, Berlin, Heidelberg, 2011.

[12] Jia, Robin, and Percy Liang. "Data recombination for neural semantic parsing." arXiv preprint *arXiv:1606.03622* (2016).

[13] Huang, Jie, Wengang Zhou, Qilin Zhang, Houqiang Li, and Weiping Li. "Video-based sign language recognition without temporal segmentation." In *Proceedings of the AAAI Conference on Artificial Intelligence,* vol. 32, no. 1. 2018.

[14] Gupta, Anvita, Alex T. Müller, Berend JH Huisman, Jens A. Fuchs, Petra Schneider, and Gisbert Schneider. "Generative recurrent networks for de novo drug design." *Molecular informatics* 37, no. 1-2 (2018): 1700111.

[15] Islam, Md Saiful, and Emam Hossain. "Foreign Exchange Currency Rate Prediction using a GRU-LSTM Hybrid Network." *Soft Computing Letters* (2020): 100009.

[16] Hochreiter, Sepp, Martin Heusel, and Klaus Obermayer. "Fast model-based protein homology detection without alignment." *Bioinformatics* 23, no. 14 (2007): 1728-1736.

[17] Kaji, Noritada, and Yoshinobu Baba. "Nanobiodevice-based single biomolecule analysis, single-cell analysis, and in vivo imaging for cancer diagnosis, cancer theranostics, and iPS cell-based regenerative medicine." *Analytical Sciences* 30, no. 9 (2014): 859-864.

[18] Sun, Jing, Attila Tárnok, and Xuantao Su. "Deep Learning-Based Single-Cell Optical Image Studies." *Cytometry Part A* 97, no. 3 (2020): 226-240.

[19] Szegedy, Christian, Vincent Vanhoucke, Sergey Ioffe, Jon Shlens, and Zbigniew Wojna. "Rethinking the inception architecture for computer vision." In *Proceedings of the IEEE conference on computer vision and pattern recognition,* pp. 2818-2826. 2016.

[20] Mirowski, Piotr W., Yann LeCun, Deepak Madhavan, and Ruben Kuzniecky. "Comparing SVM and convolutional networks for epileptic seizure prediction from intracranial EEG." In *2008 IEEE workshop on machine learning for signal processing,* pp. 244-249. IEEE, 2008.

[21] Usman, Syed Muhammad, Muhammad Usman, and Simon Fong. "Epileptic seizures prediction using machine learning methods." *Computational and mathematical methods in medicine* 2017 (2017).

[22] Eberlein, Matthias, Raphael Hildebrand, Ronald Tetzlaff, Nico Hoffmann, Levin Kuhlmann, Benjamin Brinkmann, and Jens Müller. "Convolutional neural networks for epileptic seizure prediction." In *2018 IEEE International Conference on Bioinformatics and Biomedicine (BIBM),* pp. 2577-2582. IEEE, 2018.

[23] Hoyos-Osorio, Keider, Jairo Castañeda-Gonzaiez, and Genaro Daza-Santacoloma. "Automatic epileptic seizure prediction based on scalp EEG and ECG signals." In *2016 XXI Symposium on signal processing, images and artificial vision (STSIVA),* pp. 1-7. IEEE, 2016.

[24] Osman, Ahmed Hamza, and Ahmad A. Alzahrani. "New approach for automated epileptic disease diagnosis using an integrated self-

organization map and radial basis function neural network algorithm." *IEEE Access* 7 (2018): 4741-4747.

[25] Freestone, Dean R., Philippa J. Karoly, and Mark J. Cook. "A forward-looking review of seizure prediction." *Current opinion in neurology* 30, no. 2 (2017): 167-173.

[26] Daoud, Hisham, and Magdy A. Bayoumi. "Efficient epileptic seizure prediction based on deep learning." *IEEE transactions on biomedical circuits and systems* 13, no. 5 (2019): 804-813.

[27] Sharma, Aarti, J. K. Rai, and R. P. Tewari. "Epileptic seizure prediction and identification of epileptogenic region using EEG signal." In *2015 International Conference on Green Computing and Internet of Things (ICGCIoT),* pp. 1194-1198. IEEE, 2015.

[28] Craik, Alexander, Yongtian He, and Jose L. Contreras-Vidal. "Deep learning for electroencephalogram (EEG) classification tasks: a review." *Journal of neural engineering* 16, no. 3 (2019): 031001.

[29] Ashby, Corey, Amit Bhatia, Francesco Tenore, and Jacob Vogelstein. "Low-cost electroencephalogram (EEG) based authentication." In *2011 5th International IEEE/EMBS Conference on Neural Engineering*, pp. 442-445. IEEE, 2011.

[30] Craik, Alexander, Yongtian He, and Jose L. Contreras-Vidal. "Deep learning for electroencephalogram (EEG) classification tasks: a review." *Journal of neural engineering* 16, no. 3 (2019): 031001.

[31] Lai, Chi Qin, Haidi Ibrahim, Mohd Zaid Abdullah, Jafri Malin Abdullah, Shahrel Azmin Suandi, and Azlinda Azman. "Literature survey on applications of electroencephalography (EEG)." In *AIP Conference Proceedings,* vol. 2016, no. 1, p. 020070. AIP Publishing LLC, 2018.

[32] Reilly, Colin, Patricia Atkinson, Ayesha Memon, Chloe Jones, Lyvia Dabydeen, Krishna B. Das, Christopher Gillberg, Brian GR Neville, and Rod C. Scott. "Symptoms of depression, anxiety, and stress in parents of young children with epilepsy: a case-controlled population-based study." *Epilepsy & Behavior* 80 (2018): 177-183.

[33] Zhu, Runjie, Xinhui Tu, and Jimmy Xiangji Huang. "Deep learning on information retrieval and its applications." In *Deep Learning for Data Analytics,* pp. 125-153. Academic Press, 2020.

[34] Chen, Zhenpeng, Huihan Yao, Yiling Lou, Yanbin Cao, Yuanqiang Liu, Haoyu Wang, and Xuanzhe Liu. "An empirical study on deployment faults of deep learning based mobile applications." arXiv preprint *arXiv:2101.04930* (2021).

[35] Xu, Yang, Zhixiong Li, Shuqing Wang, Weihua Li, Thompson Sarkodie-Gyan, and Shizhe Feng. "A hybrid deep-learning model for fault diagnosis of rolling bearings." *Measurement* 169 (2021): 108502.

[36] Dixit, Mayank, Abhishek Tiwari, Himanshu Pathak, and Rani Astya. "An overview of deep learning architectures, libraries and its applications areas." In *2018 International Conference on Advances in Computing, Communication Control and Networking (ICACCCN),* pp. 293-297. IEEE, 2018.

[37] Lee, June-Goo, Sanghoon Jun, Young-Won Cho, Hyunna Lee, Guk Bae Kim, Joon Beom Seo, and Namkug Kim. "Deep learning in medical imaging: general overview." *Korean journal of radiology* 18, no. 4 (2017): 570.

[38] Wang, Tonghe, Yang Lei, Yabo Fu, Jacob F. Wynne, Walter J. Curran, Tian Liu, and Xiaofeng Yang. "A review on medical imaging synthesis using deep learning and its clinical applications." *Journal of Applied Clinical Medical Physics* 22, no. 1 (2021): 11-36.

[39] Zhao, Rui, Ruqiang Yan, Zhenghua Chen, Kezhi Mao, Peng Wang, and Robert X. Gao. "Deep learning and its applications to machine health monitoring." *Mechanical Systems and Signal Processing* 115 (2019): 213-237.

[40] Wang, Jinjiang, Yulin Ma, Laibin Zhang, Robert X. Gao, and Dazhong Wu. "Deep learning for smart manufacturing: Methods and applications." *Journal of Manufacturing Systems* 48 (2018): 144-156.

[41] Plebe, Alessio, and Giorgio Grasso. "The unbearable shallow understanding of deep learning." *Minds and Machines* 29, no. 4 (2019): 515-553.

[42] Abrol, Anees, Manish Bhattarai, Alex Fedorov, Yuhui Du, Sergey Plis, Vince Calhoun, and Alzheimer's Disease Neuroimaging Initiative. "Deep residual learning for neuroimaging: An application to predict progression to alzheimer's disease." *Journal of neuroscience methods* 339 (2020): 108701.

[43] Nishi, Hidehisa, Naoya Oishi, Akira Ishii, Isao Ono, Takenori Ogura, Tadashi Sunohara, Hideo Chihara, et al. "Deep Learning–Derived High-Level Neuroimaging Features Predict Clinical Outcomes for Large Vessel Occlusion." *Stroke* 51, no. 5 (2020): 1484-1492.

[44] Zhang, Jianpeng, Yutong Xie, Qi Wu, and Yong Xia. "Medical image classification using synergic deep learning." *Medical image analysis* 54 (2019): 10-19.

[45] Panichev, Oleg, Anton Popov, and Volodymyr Kharytonov. "Patient-specific epileptic seizure prediction using correlation features." In *2015 Signal Processing Symposium (SPSympo),* pp. 1-5. IEEE, 2015.

[46] Asadi-Pooya, Ali A., and Zahra Bahrami. "Loss of responsiveness in psychogenic non-epileptic seizures." *Epileptic Disorders* 21, no. 2 (2019): 192-196.

[47] Rosas-Romero, Roberto, Edgar Guevara, Ke Peng, Dang Khoa Nguyen, Frédéric Lesage, Philippe Pouliot, and Wassim-Enrique Lima-Saad. "Prediction of epileptic seizures with convolutional neural networks and functional near-infrared spectroscopy signals." *Computers in biology and medicine* 111 (2019): 103355.

[48] Kira, Kenji, and Larry A. Rendell. "A practical approach to feature selection." In *Machine learning proceedings* 1992, pp. 249-256. Morgan Kaufmann, 1992.

[49] Guyon, Isabelle, and André Elisseeff. "An introduction to variable and feature selection." *Journal of machine learning research* 3, no. Mar (2003): 1157-1182.

[50] Visa, Sofia, Brian Ramsay, Anca L. Ralescu, and Esther Van Der Knaap. "Confusion Matrix-based Feature Selection." MAICS 710 (2011): 120-127.

[51] Stallings, William M., and Gerald M. Gillmore. "A note on "accuracy" and "precision." *Journal of Educational Measurement* 8, no. 2 (1971): 127-129.

[52] Chawla, Nitesh V., Nathalie Japkowicz, and Aleksander Kotcz. "Special issue on learning from imbalanced data sets." *ACM SIGKDD explorations newsletter* 6, no. 1 (2004): 1-6.

[53] Buckland, Michael, and Fredric Gey. "The relationship between recall and precision." *Journal of the American society for information science 45,* no. 1 (1994): 12-19.

[54] Powers, David MW. "Evaluation: from precision, recall and F-measure to ROC, informedness, markedness and correlation." arXiv preprint *arXiv:2010.16061* (2020).

[55] Fuchs, Karl H., Tom R. DeMeester, and Mario Albertucci. "Specificity and sensitivity of objective diagnosis of gastroesophageal reflux disease." *Surgery* 102, no. 4 (1987): 575-580.

In: Deep Learning and its Applications ISBN: 978-1-68507-185-1
Editor: Arvind Kumar Tiwari

Chapter 6

DEEP LEARNING AND ITS APPLICATIONS

***Palakala Sai Krishna Yadhav*[1,*], *K. Susheel Kumar*[1,†],**
***Vibhav Prakash Singh*[1,‡] *and Nagendra Pratap*[2,§]**
[1]NIT Hamirpur, Hamirpur, Himachal Pradesh, India
[2]MNNIT Allahabad, Prayagraj, Uttar Pradesh, India

Abstract

Deep learning (DL), also known as hierarchical learning or deep structured learning, is a sub-domain of machine learning. The term deep in the expression deep learning refers to the number of layers in a neural network. DL has shown a significant breakthrough in many areas with outstanding performance. After seeing the performance, DL has been used in high dimensional data and many complex applications, giving state-of-the-art results compared to conventional methods. In this paper, some applications are discussed where deep learning is widely used, namely healthcare, self-driving or autonomous cars, NLP, speech recognition, image recognition, cybersecurity, Automatic coloring, etc. Many areas are adopting deep learning. Also, we have discussed some of the DL architectures commonly used in the applications mentioned above. Deep learning is becoming more popular because of the large data available for training the networks, deep architectures, activation functions, and more computational power. And deep architectures have

*Corresponding Author's Email: 19m504@nith.ac.in.
†Corresponding Author's Email: ksusheel@nith.ac.in.
‡Corresponding Author's Email: vibhav@mnnit.ac.in.
§Corresponding Author's Email: nps@nith.ac.in.

many hidden layers that are more powerful and useful for robust feature learning to increase performance.

Keywords: Deep Learning, Convolutional Neural Network, medical images, RNN, LSTM, Neural Network

1. INTRODUCTION

A deep learning algorithm is a neural network with more than three layers, including inputs and outputs. Deep learning has recently been used in several research areas, yielding cutting-edge results. DL algorithms perform admirably on large amounts of data, whether labeled or unlabeled. Nowadays, a large amount of data is available, both supervised and unsupervised, in audio, CSV files, images, and text. This large amount of data can be used to train DL algorithms, which improve model performance. Using the conventional method, predictions in data and feature extraction is a challenging task [1]. In this deep learning era, the above scenario has changed, as the DL algorithms can learn from the data automatically, like finding the patterns in data using many hidden layers. And more hidden layers are used to understand the data at a high level and more effectively. Earlier problems in research areas that took a lot of time for feature extraction, preprocessing, and data predictions are easily solved using DL algorithms with the help of computation power. The computational resources available in today's world are one of the biggest advantages of deep learning. Because of the available computational resources, the models' time is less than the traditional ML approach. Because a large amount of data is available, the depth in the neural network models has increased significantly, allowing for more data abstraction such as patterns, edges, and features from the model's input. Another advantage of DL algorithms is that they can learn from unlabeled data and only require a large dataset. If the dataset is small, then data augmentation can be done to increase the size of the dataset. After observing an improvement in the performance of research projects using DL algorithms. Now many research areas started adapting deep learning in many areas such as medical imaging, voice assistants, self-driving cars, image recognition, fraud detection, advertising, finance, etc. For example, medical imaging predicts whether a patient's cardiovascular disease is normal or abnormal.

Similarly, in healthcare, many critical diseases can be predicted or detected early using medical images such as X-rays, CT scans, MRI scans, and so on. Recently, DL algorithms detected the outbreak of the covid19 pandemic using CT scans of patients. According to Zhang et al. [19], another example of deep learning in 5G and mobile networks, mobile devices are increasing exponentially, and processing data between mobile and network is a difficult task with low battery utilization. DL is being used to address these issues and improve the performance of mobile devices. Barz et al. [2] proposed a new approach for training DL algorithms on small datasets. The data given to DL algorithms are typically very large, but the algorithms do not perform well with small datasets. Obtaining data in some research areas is indeed costly. According to paper [2] changing the loss function from categorical to cosine loss improves model performance for small datasets.

2. Overview of Architectures

2.1. Supervised Architectures

Supervised learning is a method in which the data has been given to the model or algorithm is clearly labeled, and the output of the model or algorithm is classified as correct or not. Here the dataset is split in a ratio required for training the model for the prediction, the validation set is used to check the performance of the model and try to improve it as high as possible, and the remaining dataset is used for the evaluation of the algorithm or model in terms of prediction and effectiveness. The metrics used to evaluate any supervised algorithms are the confusion matrix, root mean squared error and others. The most important metric for classification is accuracy [6]. There are two types of supervised algorithms: classification and regression. These algorithms are used for dataset predictions and work on labeled data. The predicted outputs in classification are discrete values such as binary classes like 0 or 1, multiple binary classes like play or not play, cat or not a cat, dog or not dog, and multi-class if more than three classes [6]. One example of a multi-class classification is the MNIST dataset. The images are handwritten numbers, and the objective is to predict the numbers from 0 to 9, resulting in a 10-class classification. An example of binary classification is spam emails, which classify the output as spam or not. Accuracy is the most commonly used metric for classification problems.

In regression algorithms, the predicted outputs are continuous values such as scored labels [6]. One of the differences between both algorithms is, classification is used to predict the set of classes, and regression is used to predict the quantity. One example is weather prediction, where the model is trained on previous weather data and predicts future weather days. For regression algorithms or models, the most common metric is RMSE.

2.1.1. Convolutional Neural Networks

CNN is a multilayered neural network, which is particularly used for images. Usually, Neural networks aren't built to extract features from images and aren't capable of doing. So, convolutions and pooling are used to extract features from the images, and they can't perform classification so, we need fully connected layers to classify the data.

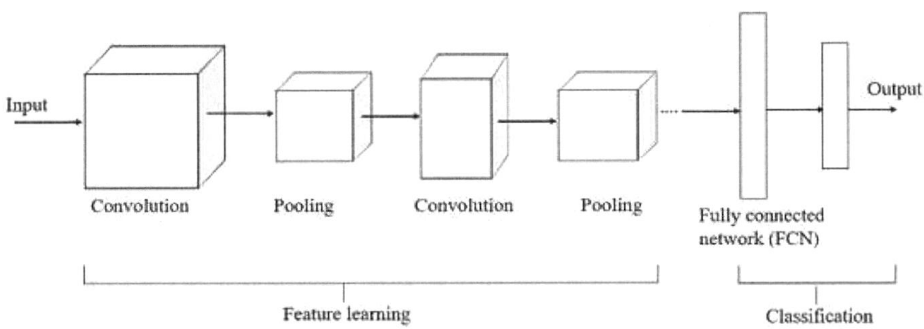

Figure 1. Convolutional Neural Network.

Fig. 1 shows multiple layers of convolutions and pooling, and the more layers there are, the better the algorithm, and these layers are used to extract features from images. Pooling is used to reduce the dimensionality of features. Max pooling is commonly used because it selects the most information from the extracted features. Following that, we used to flatten, softmax, and FCN layers; we can use dense layers in a regression model. The most difficult challenge in traditional ML algorithms is feature extraction from images, but CNN automatically extracts features and learns from them.

2.1.2. Recurrent Neural Networks

Assume the objective is to predict the letter or word in a sentence. Typically, in CNN, there are many hidden layers, and each layer is independent so, the weights and activations are different for each layer. Now using CNN to predict the next word in a sequence is difficult because the layers are independent. All the hidden layers are combined in the recurrent network as they have the same bias and weights. Lets say we have the sentence nature is beautiful in this sentence. After giving the first three words as input to the RNN, it must predict the fourth word called beautiful. To predict the fourth word, it has to maintain the memory of the previous inputs. So, in RNN, all the hidden layers are rolled and combined.

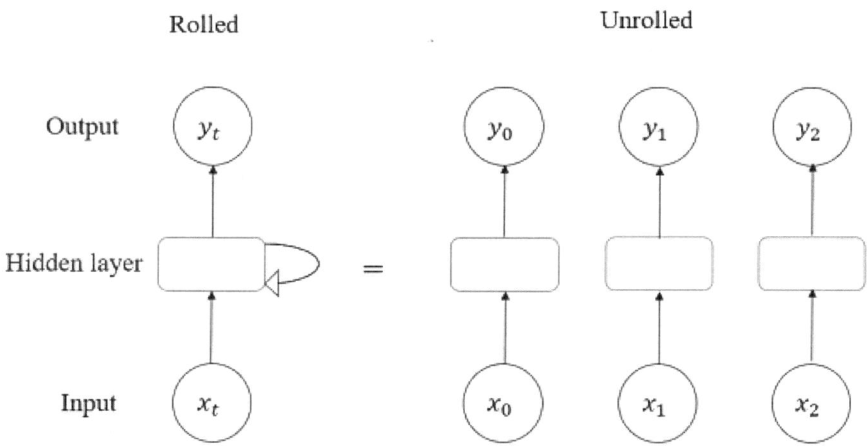

Figure 2. Recurrent Neural Network.

2.1.3. Long Short-Term Memory

As shown in fig 3, RNN can predict the next word of the sentence or the next letter in a word. It can predict only if the sentences or the data input to the network is small. If the large corpus is given as input to the network, it cannot predict the sequence word for that sentence. Because RNN saves the input data for the short term, it means if ten lines of text data are given input, it will be kept on replacing the existing data because it doesn't know which word is important and which word is not important. Using LSTM, we can overcome the above problem. LSTM has a memory cell that remembers the important data and saves it in a memory cell for a long time. The memory cell has three gates that are useful for the prediction of words even with large inputs. The input gate in the cell will maintain what information needs to be sent or not; the output gates will maintain when to send the outputs from the cell. The forget gate will decide which information is important and what to store and what to forget in the input data. As shown in Fig. 3, the cell will maintain the weights to control these gates.

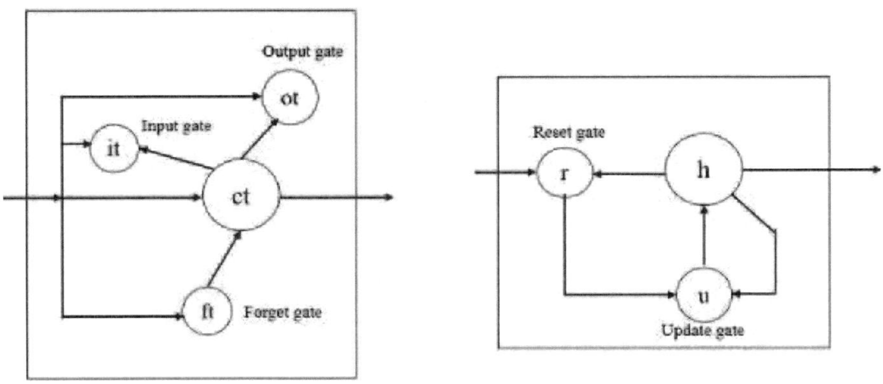

Figure 3. LSTM and GRU.

2.1.4. Gated Recurrent Unit

GRU is similar to LSTMs in that it is very easy to train and simpler than LSTM. It has an update gate that maintains the contents of the previous cell and a reset gate that determines how much information to forget in previous cells.

2.2. Unsupervised Architectures

In Unsupervised learning, the data provided to the model is unlabeled. Here the goal is to predict the patterns or structure in the dataset and divide the data into similar groups to get some insight into the data, also known as clustering. And understanding the data, how it is distributed in space is known as density estimation. And autoencoders are used for ignoring the noise in data, typically for the dimensionality reduction in the unlabeled data. Usually, these methods are applied when the dataset is very large and too expensive to label the data. Then deep learning comes into handy for these types of problems in the unlabeled data.

2.2.1. Autoencoders

Autoencoders are an unsupervised learning method used for data encoding and decoding; it is also known as the ANN. Autoencoders are popular for the bottleneck layer in the network, which can be built by using encoding functions. It has three layers; the first layer is the input layer which takes the input and passes to the bottleneck layer or hidden layer; the second layer is the bottleneck layer or hidden layer; as shown in Fig. 4, the hidden layer is significantly smaller compared to the input layer. Because the encoder function is used for the compressed representation of data, which forces the network to remove the noise or redundant information. The third layer is the output layer using the decoder function. It reconstructs the input layer. And finally, an error function is used to compute the difference between the input and output layers.

2.2.2. Restricted Boltzmann Machines

RBM is a two-layered neural network, and it is similar to the Boltzmann machine. RBM has two layers. One is the visible layer or input layer, and the other is the hidden layer. Here, all the neurons in the visible layer are connected to every node in the hidden layer in a bipartite graph manner. But in traditional BM, it is symmetrically connected like all nodes in the visible layer are connected within the layer and every node in the hidden layer. During the training, the probability distribution is calculated using a stochastic approach. And each layer has a bias, the visible bias is used to reconstruct the input layer, and hidden bias is used to build the activation in the forward pass. RBMs are mostly used for dimensionality reduction.

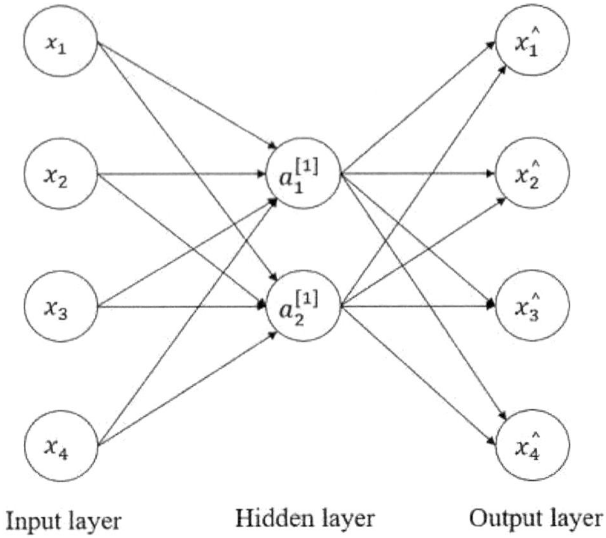

Figure 4. Autoencoders.

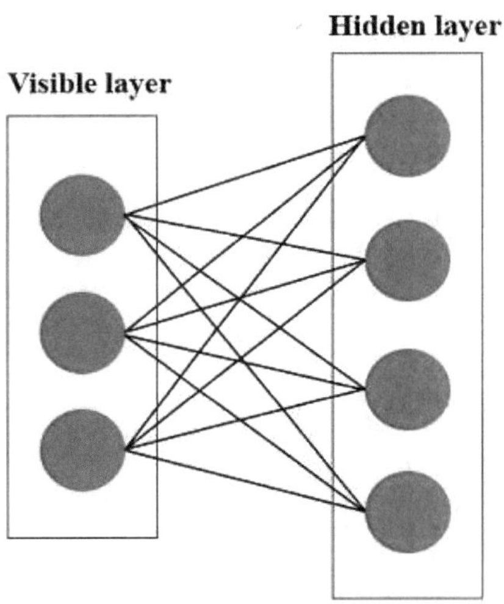

Figure 5. Restricted Boltzmann Machine.

2.3. Deep Belief Networks

DBN is a deep multilayer network architecture, which has many hidden layers. And in DBN, each pair of linked layers is an RBM. Here the training phase is done using an unsupervised manner, and fine-tuning is done using supervised. In the training phase, the input layer takes the input from the data and passes it to the first hidden layer; the pair of these two layers is an RBM. Similarly, the first hidden layer acts as input to the second hidden layer, and the procedure continues till the model is trained. In the output layer, the network classification is done for labeling the nodes in the layer. After applying the backpropagation or gradient descent to the network, the training process is complete.

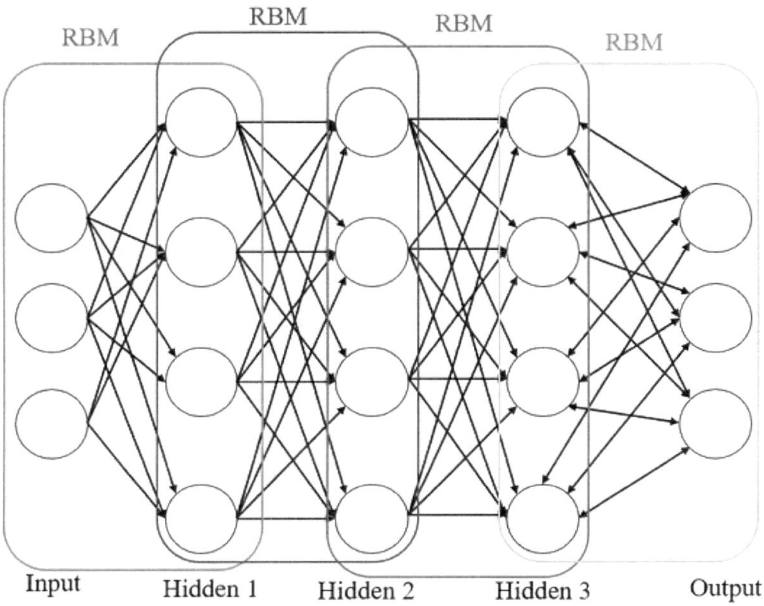

Figure 6. Deep Belief Network.

3. Applications

The following are some applications of deep learning which are growing very fast recently.

3.1. Natural Language Processing

NLP is used to analyze the text in a computerized approach. The main goal is to understand and process the human language. Even though many types of research are going on actively, the goal of NLP is not yet achieved. In classical NLP, most work is done manually, like performing preprocessing such as tokenization, lemmatization, removing stop words, etc. The next feature extraction is done, and some models or algorithms are applied along with inference and output. All these steps are done manually, and there might be some mistakes or missing out on some good features that can happen while performing manually. Coming to NLP using DL, the most challenging task like selecting features and preprocessing is done automatically as it learns features independently. DL algorithms filter out the best features and try to increase the performance at the highest level. The response time of DL algorithms is very less compared to the traditional or classical NLP algorithms. The advantages of using DL over classical NLP are getting the highest level of accuracy, doing challenging tasks on its own, and responding with very little time.

Araque et al. [1] proposed a novel approach integrating a traditional model with the DL technique, giving a better F1 score than the deep learning technique. First, they have built a DL algorithm with word embeddings and next built the two-ensemble technique, which is mostly used in NLP and later combined both models and has taken the best feature compared to the deep learning technique. Majumder et al. [12] developed a CNN model in which text essays are input to extract personality traits from the documents. They proposed five networks, all of which are binary classifiers, the input text corpus is converted to n-gram vectors. Finally, the model predicts whether the traits are positive or negative; these vectors are given to FCN, which classifies the document to either one of the traits.

3.2. Healthcare

Deep learning is reshaping the healthcare industry by developing new possibilities to improve people's lives. DL is used in medical imaging, computer-aided disease detection, genome analysis, discovering new drugs, etc. Many diseases have been identified using DL, like finding a tumor in the brain, identifying the blockage in the heart, vessel blockage in the retinal, and finding the abnormalities using x-rays, CT scans, and

MRI images. DL is also applied in IoT devices to monitor the ECG, heart rate, and medical diagnosis [17]. Khened, M et al. [9] developed a CNN model with the ensemble of classifiers to detect cardiac diseases using the volume estimation of the left ventricle. Their algorithms have shown 100% accuracy in the classification of diseases. The images taken are in Dicom format and a short-axis view of the heart. In this paper, DL is used to detect cardiac diseases using MRI images. Jindal V et al. [5] applied DL in wearable devices for monitoring the real-time data. Applying the DL on low-power devices is a challenging task; the proposed design in this paper can overcome the above limitation. They have combined external network and inertial sensor data features, and they have used spectral-domain preprocessing to optimize DL on the devices.

3.3. Image Recognition

CNN is extensively used in the image and video recognition area. In recent years, the growth in image processing using DL has been increased a lot due to the extensive results in the classifications. The widely used DL algorithm for images is CNN, it has multiple layers of convolutions and multiple pooling layers, which are used to reduce the size of images and extract the best features from the image, and after learning the features, the results are fed to the FCN which classifies the images. This image recognition process is applied in many areas like medical, mineral detection, autonomous and biometric, etc. Srinivasa K et al. [18] applied CNN and LSTM to detect the expressions in the video or images. In todays world, a massive amount of multimedia data is available; this paper aims to detect facial expressions in an image or frame. For detecting the expressions, the movement of facial muscles and eye blinking is also important. Traditional ML methods such as KNN, SVM, K- means are already implemented, but they cannot process the large data and capture each frame in a video and aggregate the output. Using LSTM and CNN, it can capture each frame in a video and analyze the expression with a range between 1 100. Finally, it aggregates all the frames using the abovementioned DL algorithms, and a large amount of data can also be handled. Iliadis M et al. [8] proposed a framework for recovering the video frames using deep learning. Using DL, the quality of reconstructing the video frames has significantly improved compared to traditional ML methods. And the frames are restored in a matter of seconds using the method described. After adding more layers to the network, the performance also significantly improved.

3.4. Autonomous or Self-Driving Cars

In the automobile industry [16], deep learning is forced into the self-driving car. Many research in self-driving cars has been significantly increasing. The key issues to build a self-driving car are the dataset and the network. A massive amount of data is required to train the DL model, consisting of cars, pedestrians, footpaths, etc.; all these data cannot be found in a single dataset. Even after collecting such a massive dataset, there are still cases like damaged roads, accidents, etc., which are very hard to solve. The next issue is a neural network; after building the model with many deep layers, getting a high-level accuracy is challenging. Even if the training model is 100% accurate, there is no guarantee that no errors will occur. Maqueda A et al. [13] have used event cameras that capture dynamic vision without redundant information. Event cameras capture the moving edges in a frame; the images are low-latency and high dynamic range, which is an advantage compared to traditional cameras. Using, these images CNN is applied to predict the steering angle of the vehicle. Similarly, Ramos et al. [15] proposed an obstacle detection framework using DL to detect small road hazards. In this paper, the proposed FCN network predicts the obstacles of even 5cm height on the road, and the network's performance has significantly improved.

3.5. Cybersecurity

Since the number of treats in the network has risen dramatically, several companies worldwide are looking for new solutions to reduce the number of treats in internet-connected devices. These treats can be detected using conventional methods like host security and network security systems. The security models used to detect the threats are firewalls, antiviruses, intrusion detection systems, etc. Using DL, the time taken to analyze and identify the features in the malware is reduced. DL algorithms can identify the features in the malware attacks in a robust way. The use of DL has improved the detection of new types of network attacks as it is done automatically and saves a lot of time in feature engineering. [11] Dali Zhu et al. [3] say that Android is open-source, allowing many hackers to create new malware that cannot be detected using traditional ML methods. In this paper, deep flow, a novel DL method, is used to access the data directly from the applications that detect the malware. The proposed method has given a high F1 score compared to the conventional methods.

3.6. Speech Recognition

Neural networks applied to speech recognition in the early '90s, which was popular back then but couldn't outperform the Gaussian mixture models (GMMs) [4]. Lately, deep neural networks (DNN) became popular because of the availability of large training data, new architectures of DNN, computing power, and activation functions. The deeper the network, the powerful the model, but the training was slow even with GPUs. Alternatively, RBMs can also be used for training the data. Huang J et al. [7] applied DBN to the audio-visual speech recognition (AVSR) over traditional GMM and Hidden Markov model (HMM) methods. The features extracted using DBN show better results compared to GMM/HMM. So, they proposed a method that combines features of AVSR using DBN with GMM has shown significant results, which reduced the error rate by 21%. Noda K et al. [14] proposed an integrating HMM model. First, the audio features are extracted using autoencoders, and after pre-processing, the features are denoised. Secondly, the visual features are extracted using CNN, which gives robust features. And in the end, these acquired features are integrated using HMM.

3.7. Automatic Coloring

Coloring the grayscale images is a challenging task. It can be done manually with human intervention, which is a time taking and difficult task. DL can be used to colorize the black and white images by using the context and objects in the image. This can be done using the pre-trained model on the ImageNet dataset, a very large and high-quality dataset, and supervised layers are added to recreate the image in color. The colorization using DL algorithms is visually impressive and very fast compared to the conventional methods. Using this approach, the frames of the black and white movies can also be colored. Larsson G et al. [10] developed a fully automatic DL network for coloring the grayscale images. Using DL, coloring the images is less time-consuming and cost-efficient. The proposed models predict per pixel color histogram, which gives a color image as output automatically.

Conclusion

An overview of deep learning is provided in this paper. Deep learning is growing fast in many areas. Researchers widely adopt it with a significant increase in performance because of large data available for training, deeper architectures, more computing power, and many libraries available for the DL techniques. One of DL's major advantages is that it automatically learns the robust features of the data, which are very hard to retrieve using traditional methods. In this paper, a brief explanation of widely used networks and algorithms is mentioned. Many applications have been using deep learning, and some of the applications are mentioned in this paper. Deep learning has a lot of potentials to grow in many areas because of less human intervention.

References

[1] Araque, O., Corcuera-Platas, I., Snchez-Rada, J.F., Iglesias, C.A., 2017. Enhancing deep learning sentiment analysis with ensemble techniques in social applications. *Expert Systems with Applications* 77, 236–246. https://www.sciencedirect.com/science/article/pii/S0957417417300751, doi:https://doi.org/10.1016/j.eswa.2017.02.002.

[2] Barz, B., Denzler, J., 2020. Deep learning on small datasets without pre-training using cosine loss, in: *2020 IEEE Winter Conference on Applications of Computer Vision (WACV)*, pp. 1360–1369. doi:10.1109/WACV45572.2020.9093286.

[3] Dali Zhu, Hao Jin, Ying Yang, Wu, D., Weiyi Chen, 2017. Deepflow: Deep learning-based malware detection by mining android application for abnormal usage of sensitive data, in: *2017 IEEE Symposium on Computers and Communications (ISCC)*, pp. 438–443. doi:10.1109/ISCC.2017.8024568.

[4] Deng, L., Hinton, G., Kingsbury, B., 2013. New types of deep neural network learning for speech recognition and related applications: an overview, in: *2013 IEEE International Conference on Acoustics, Speech and Signal Processing*, pp. 8599–8603. doi:10.1109/ICASSP.2013.6639344.

[5] Dixit, M., Tiwari, A., Pathak, H., Astya, R., 2018. An overview of deep learning architectures, libraries and its applications areas, in: *2018 International Conference on Advances in Computing, Communication Control and Networking (ICACCCN)*, pp. 293–297. doi:10.1109/ICACCCN.2018.8748442.

[6] Hatcher, W.G., Yu, W., 2018. A survey of deep learning: Platforms, applications and emerging research trends. *IEEE Access* 6, 24411–24432. doi:10.1109/ACCESS.2018.2830661.

[7] Huang, J., Kingsbury, B., 2013. Audio-visual deep learning for noise robust speech recognition, in: *2013 IEEE International Conference on Acoustics, Speech and Signal Processing*, pp. 7596–7599. doi:10.1109/ICASSP.2013.6639140.

[8] Iliadis, M., Spinoulas, L., Katsaggelos, A.K., 2018. Deep fully-connected networks for video compressive sensing. *Digital Signal Processing* 72, 9–18. URL: https://www.sciencedirect.com/science/article/pii/S1051200417302130, doi:https://doi.org/10.1016/j.dsp.2017.09.010.

[9] Khened, M., Kollerathu, V.A., Krishnamurthi, G., 2019. Fully convolutional multi-scale residual densenets for cardiac segmentation and automated cardiac diagnosis using ensemble of classifiers. *Medical Image Analysis* 51, 21–45. URL: https://www.sciencedirect.com/science/article/pii/S136184151830848X, doi:https://doi.org/10.1016/j.media.2018.10.004.

[10] Larsson, G., Maire, M., Shakhnarovich, G., 2016. Learning representations for automatic colorization, in: Leibe, B., Matas, J., Sebe, N., Welling, M. (Eds.), *Computer Vision – ECCV 2016*, Springer International Publishing, Cham. pp. 577–593.

[11] Mahdavifar, S., Ghorbani, A.A., 2019. Application of deep learning to cybersecurity: A survey. *Neurocomputing* 347, 149–176. URL: https://www.sciencedirect.com/science/article/pii/S0925231219302954, doi:https://doi.org/10.1016/j.neucom.2019.02.056.

[12] Majumder, N., Poria, S., Gelbukh, A., Cambria, E., 2017. Deep learning-based document modeling for personality detection from text. *IEEE Intelligent Systems* 32, 74–79. doi:10.1109/MIS.2017.23.

[13] Maqueda, A.I., Loquercio, A., Gallego, G., Garca, N., Scaramuzza, D., 2018. Event-based vision meets deep learning on steering prediction for self-driving cars, in: *2018 IEEE/CVF Conference on Computer Vision and Pattern Recognition*, pp. 5419–5427. doi:10.1109/CVPR.2018.00568.

[14] Noda K, Yamaguchi Y, N.K., 2015. Audio-visual speech recognition using deep learning. *Applied Intelligence* volume doi:10.1007/s10489-014-0629-7.

[15] Ramos, S., Gehrig, S., Pinggera, P., Franke, U., Rother, C., 2017. Detecting unexpected obstacles for self-driving cars: Fusing deep learning and geometric modeling, in: *2017 IEEE Intelligent Vehicles Symposium (IV)*, pp. 1025–1032. doi:10.1109/IVS.2017.7995849.

[16] Rao, Q., J., F., 2018. Deep learning for self-driving cars. *1st International Workshop On Software Engineering For AI In Autonomous Systems.* doi:10.1145/3194085.3194087.

[17] Rav, D., Wong, C., Lo, B., Yang, G., 2017. A deep learning approach to on-node sensor data analytics for mobile or wearable devices. *IEEE Journal of Biomedical and Health Informatics* 21, 56–64. doi:10.1109/JBHI.2016.2633287.

[18] Srinivasa, K.G., A.S., 2018. Performance analysis and application of expressiveness detection on facial expression videos using deep learning techniques. *Data-Enabled Discov* 9.

[19] Zhang, C., Patras, P., Haddadi, H., 2019. Deep learning in mobile and wireless networking: A survey. *IEEE Communications Surveys Tutorials* 21, 2224–2287. doi:10.1109/COMST.2019.2904897.

In: Deep Learning and its Applications
Editor: Arvind Kumar Tiwari
ISBN: 978-1-68507-185-1

Chapter 7

AN INTRODUCTION TO SENTIMENT ANALYSIS USING DEEP LEARNING TECHNIQUES

Rajnish Pandey[1,*], Jyoti Prakash Singh[1] and Sudhakar Tripathi[2]
[1]NIT, Patna, Bihar, India
[2]REC, Ambedkarnagar, UP, India

ABSTRACT

Examination of public opinion (also called opinion mining or sentiment analysis (SA)) is a hot topic in the natural language processing (NLP) field. The SA aims to define, extract, and organize sentiments from user-generated texts in places like social media, blogs, news headlines, and reviews on product. Most of the researchers in the literature have used machine learning methods to address SA tasks from various perspectives over the last two decades. Since an NLP researchers' output is heavily influenced by data representation choices, many studies focus on developing powerful feature extractors using

[*] Corresponding Author's Email: rajnishp.phd19.cs@nitp.ac.in.

domain knowledge and careful engineering. Deep learning techniques have recently emerged as effective computational frameworks capable of automatically discovering complex semantic representations of texts from data without the need for function engineering. Many SA activities, such as sentiment recognition, opinion extraction, fine-grained SA, and others, have benefited from these methods. In this work, we provide an overview of effective deep learning approaches to SA activities at various scales.

Keywords: natural language processing, sentiment analysis, opinion mining

1. Introduction

SA is a branch of psychology [Pang & Lee (2006)] that analyses people's thoughts, feelings, and emotions derived through customer script automatically. Sentiment analysis is a hot topic in natural language processing, and it's still getting a lot of attention in data mining, because emotions are powerful drivers of social behavior [Batrinca & Treleaven (2015)]. With the exponential development of social networking platforms like Twitter and Facebook, as well as review sites like IMDB, Amazon, and Yelp, sentiment analysis is gaining traction in the academic and business communities [Alsaeedi & Khan (2019)].

Let's use an analogy to illustrate what "sentiment" means. Assume a user wrote, "I purchased a OnePlus yesterday. It's a fantastic phone. The touch screen is truly incredible. However, for middle-class families, the price is little high". Various users can refer to this review in order to gain some insight into the above-mentioned product. The users may or may not purchase the above cellphone based on this review [Berka (2020)].

Sentiment analysis tries to find all the sentiment quintuples in a textual material, according to the concept of sentiment. The five elements of the sentiment quintuple are used to create sentiment analysis activities [Manning & Schutze (1999)]. Sentiment grouping at the document/ sentence level, for example, focuses on the third dimension (neutral, negative, and positive sentiments) while avoiding remaining elements.

The first four dimensions of the quintuple are the subject of fine-grained opinion extraction. The second and third dimensions are the subject of target-dependent emotion grouping [Pang et al. (2002) Pang, Lee & Vaithyanathan].

Machine learning-based approaches have dominated most emotion analysis activities over the last two decades [Alswaidan & Menai (2020)]. Since feature representation has such a large impact on machine learning efficiency, several researches in the published studies concentrate upon using successful characteristics in conjunction with field knowledge and cautious engineering. Representation learning algorithms, on the other hand, will prevent this by automatically discovering discriminative and explanatory representation of text from input. Deep learning is a class of description learning method in which nonlinear neural networks have been used to learn several levels of representation, including the one that converts the transformation of a reflection from one point to a higher point and represents in more abstract manner [Zhang et al. (2018) Zhang, Wang & Liu]. It is possible to portray can be added to identification and classification tasks as a function. We present successful deep learning algorithms for sentiment analysis in this chapter. In this part, the term "deep learning" includes the application of neural network techniques to automatically learn text features that are both real-valued and continuous from the textual dataset.

Deep learning is indeed a subset of machine learning that deals with artificial neural networks, which are systematic procedure induced by the structure and operation of the brain [Zador (2019)]. You may be perplexed either you are new to the aspect of deep learning or if you plenty of experience with neural networks. Many of our friends and others who studied and in the mid-1990s, many people, especially younger, utilized neural networks, were initially puzzled.

This chapter is structured as follows. We begin by discussing the learning methods. Word embedding is another term for consistent word vector representation since words are the fundamental computational unit of natural language processing. These term embeddings will be used as inputs for sentiment analysis activities in the future. We characterize linguistic compositional approaches for statement emotion detection, for fine-grained perception extraction, neural serial models have been used first. Finally, we wrap up this chapter with some recommendations for the future.

2. EMBEDDINGS

The aim of word representation is to reflect different facets of a word's context. For in- stance, a "cellphone" may be represented in a variety of ways, along with information such as that cellphones are mobile devices with a battery, processors, network IC and display. Encoding a word as a single-hot vector is a simplest way of embedding. Both the one-hot vector and vocabulary were of the same length, except only one of the dimensions is 1, while the rest are all 0. The one-hot word representation, on the other hand, just encodes the indices of terms in a language, failing to capture the lexicon's rich contextual structure.

It is often observed that learning word clusters for discovering word similarity of different domain plays a very important role in NLP. Each word based on some common features belongs to a distinct class, and words belonging to the same class are identical in certain ways [Das et al. (2017) Das, Ganguly & Garain]. As a output, in a one-hot representation we can achieve smaller vocabulary number. Many researchers aim to learn a real-valued and continuous vector for each word, also termed by word embedding, rather than describing correlation with even a data type based on clustering findings that lead to a soft or hard partition of the set of terms. The distributional hypothesis notes that terms in common ways have similar meanings. The majority of current embedding neural networks are built on this hypothesis [Ruas et al. (2020) Ruas, Ferreira, Grosky, de França & de Medeiros].

To represent each text into its corresponding vectors is a necessary step to provide input to the deep learning models. In this chapter, we have used word embedding and character embedding both for extracting the robust features from the text. For word embedding vectors, we can tokenize each sentence into words, and then we can create a n-dimensional embedding vector for each word [Young et al. (2018) Young, Hazarika, Poria & Cambria]. Often word embedding vectors are initialized randomly from the uniform distribution. These weights are then trained during the back-propagation. It is found that irrespective of word embedding vectors, character embedding vectors are achieving prominent performance for the social media text which containing several grammatical mistakes, spelling mistakes, and non-standard abbreviations [Arora & Kansal (2019)]. Persona based methods have the advantage of having a limited vocabulary as well as the ability to handle any sentences, syntax, or even other grammar rules. This comes as a result of bigger models that take longer to practice. Therefore, along with the word embedding vectors, utilize character embedding vectors to give input to the hybrid convolutional neural network for the classification of the statements.

Matrices factorization approaches may be interpreted as modelling word representations in order to achieve this aim. Word embeddings are often learned through term co-occurrence estimates using the Hellinger PCA form [Almeida & Xexéo (2019)]. Since they don't provide task-specific detail, it's question-able if conventional matrix factorization methods are appropriate for a goal purpose. This issue is addressed by Supervised Semantic Indexing, which considers the super- vised knowledge of a given task (e.g., information retrieval). With a margin rating loss, they derive the principle of individual via tap data results. In information retrieval, DSSM may be thought of as studying with no supervision, task specific textual embeddings are developed.

The fundamental assumption behind sentiment-specific strategy [Jianqiang et al. (2018) Jianqiang, Xiaolin & Xuejun] is that if a word sequence's gold sentiment polarity is positive, positive score that should be expected must be better than the negative score.

3. Sentiment Classification at the Sentence Level

Sentence-level emotion analysis is concerned with categorizing the polarities of a sentence. We usually split the polarities of one sentence $w_1w_2...w_n$ into two (+/-) or three (+/-/0) groups, where -, + and 0 represent negative, positive and neutral respectively. A representative sentence classification issue is the assignment. Emotion recognition at the sentence label is possible through two-phase system using a neural network arrangement, one of which is a module for representing sentences that uses complex neural architectures and the other is based on softmax operation for classification [Gu et al. (2018) Gu, Wang, Kuen, Ma, Shahroudy, Shuai, Liu, Wang, Wang, Cai et al.].

Using word embeddings and pooling techniques for each sentinel word, anyone can get a basic exemplification for a textual statement. To extract the essential features from the textual sentence pooling operation is exhibited [Chen et al. (2018) Chen, Ling & Zhu].

To check their suggested sentiment-encoded word embeddings [Jianqiang et al. (2018) Jianqiang, Xiaolin & Xuejun] three pooling approaches. The approach is also one example of how sentences can be represented. Recent developments in phrase presentation in order to classify sentences, in particular, go far above this. In the literature, several complex neural network architectures have been suggested. We clustered similar research into four class: (1) convolutional neural networks, (2) recursive neural networks, (3) recurrent neural networks, and (4) improved sentence representation using auxiliary tools. In the subsections, we will present these works.

3.1. Convolutional Neural Networks for Textual Dataset

Basic idea of a convolutional neural network (CNN) is derived from the various mathematical concepts such as (i) convolution, (ii) pooling,

(iii) flatten, and (iv) fully connected network in the context of textual dataset [Jacovi et al. (2018) Jacovi, Shalom & Goldberg]. Based on filter specification, convolution will generate textual features. These features were then passed to pooling operation for further reduction of the features depending upon the window size of pooling operation. The obtained features are arranged into a single column vector and will work as the input to a fully connected network. This conversion is also known as flattening operation. Based on these operations any statement can be classified into different classes.

A convolution layer traverses a sequential input with a constant-size filter to perform transformations in nonlinear manner. If we provide an input sequence of $p_1p_2...p_n$ and assume the size of the local filter is K, we can get a concatenated output of $o_1o_2...o_{n-K+1}$:

$$o_i = f\left(\sum_{k=1}^{K} w_k p_{i+k-K}\right) \tag{1}$$

The tanh(.) and sigmoid(.) are examples of activation functions. Where K = n (for e.g., 3) and x_i is the input word embedding, o_i is a nonlinear mixture of p_i, p_{i+1}, and p_{i+2}, similar to the functions for mixed unigrams, bigrams, and trigrams, which concatenate the surface types of the respective terms in a difficult way.

[Kalchbrenner et al. (2017) Kalchbrenner, Grefenstette & Blunsom] improve the simple CNN model in two ways for improved sentence representation. In the one contrary, they employ complex k-max pooling, which reserves while pooling, the highest- k values rather than just one value for each dimension as in basic max pooling. The meaning k is dynamically determined by the duration of the sentence. They, on the other hand, use multilayer CNN architectures to increase the number of layers in a CNN, centered on the assumption that the sophisticated attributes can be encoded by deep neural networks.

To best reflect words, some CNN variants have been investigated. The operator for variational, nonconsecutive computation suggested by [Li et al. (2018) Li, Liu, Chen & Rudin] is one of the most representative works. The operator uses tensor algebra to obtain all n-word pairs, regardless of whether the terms are in succession. The procedure is repeated, starting with one letter, then two-word, and finally three-word combos. The following equations are used to derive both unigram, bigram, and trigram features:

$$z_i^1 = Ap_i \tag{2}$$

$$z_i^2 = s_{i-1}^1 oBp_i \tag{3}$$

$$z_i^3 = s_{i-1}^2 oCp_i \tag{4}$$

where $s_i^1 = \mu s_{i-1}^1 + z_i^1$ and $s_i^2 = \mu s_{i-1}^2 + z_i^2$

A, *B*, and *C* are model specification, μ is a hyper-parameter, and *odot* stands for element-by-element product. Finally, they put together combinations of the three types of features to form a sentence representation.

The exploration of heterogeneous input word embeddings has been the subject of a number of studies. [Kim (2019)], for example, investigates three related approaches to word embedding. The author considers the effect of dynamic fine-tuning on two separate embeddings, a randomly initialized embedding and a pretrained embedding, using two different embeddings. Eventually, it suggests that CNNs with several channels, depends on diverse embeddings, which incorporates the two types of embeddings. [Yin & Schtze (2015)] build on this work by using multichannel multilayer CNNs to incorporate several separate word embeddings. Further- more, for the model weight initialization, they use detailed pretraining techniques. [Gu et al. (2018) Gu, Wang, Kuen, Ma, Shahroudy, Shuai, Liu, Wang, Wang, Cai et al.] propose a condensed version of it, which demonstrates improved results in the mean- time.

Term embeddings may also be used to improve word representation by using character-level functionality. In essence, a neural network always had to create representation of words through input character fragments in the same manner used for building word vectors. As a result, we can derive word representations by applying a regular CNN form to the character embedding sequences. The consequence of such an expansion is investigated by [Schmidt et al. (2019) Schmidt, Marques, Botti & Marques]. The final word representations for sentence encoding can be improved by concatenating the term of descriptions at the character stage using the original embedding.

3.2. Recurrent Neural Networks for Textual Dataset

The CNN normally held the characteristics of the local composition in the vicinity of a certain area using a fixed-size word window, yielding encouraging results. It, but at the other hand, ignores features of long-distance dependence that represent syntactic and semantic knowledge, which are crucial in comprehending linguistic analysis. The re-current neural network (RNN) in the neural setting is used to solve these dependency-based functions, with considerable results. A typical RNN evaluates the output secret vectors in a timely order manner. $v_i = f(Wp_i + Uo_{i1} + b)$, where p_i denotes the input vector. From above representation, we can analyze o_i, the current output that depends not only on the current input p_i, but even on the previous performance that was secret v_{i1}. In either way, the new secret output will provide unrestricted relations to previous input and output vectors.

[Wang et al. (2015) Wang, Liu, Sun, Wang & Wang] present the first study of tweet sentiment processing using long short-term memory (LSTM) neural networks. They start by applying a state-of-the-art RNN to a sequence of input embeddings. $p_1p_2...p_n$, and use the last secret output v_n to represent the final representation of a single sentence. The authors then propose using an LSTM-RNN structure instead of a standard RNN because, gradient collapse can affect normal RNNs and diminishing returns, while LSTM is much simpler because it uses three gates, as well as a storage unit to bind input and output vectors.

3.3. Recursive Neural Networks for Textual Dataset

The use of a recursive neural network to simulate the structural inputs to trees generated by explicit syntactic parsers was recently proposed. [Socher et al. (2013) Socher, Perelygin, Wu, Manning, Ng & Potts] propose a matrix-vector neural network based on recursion that combines two leaf nodes to have an illustration of its source node. Each sentence formulation is therefore built iteratively from the roots up. They start by converting the input holding trees into a binarized tree, which has two leaf nodes for each parent node. The binary tree is then subjected to a recursive algorithm using function engineering.

In addition, [Socher et al. (2013) Socher, Perelygin, Wu, Manning, Ng & Potts] use low-rank tensor procedures to replace vector recursion using $v_p = f\,(v_l T v_r)$ to calculate the image between root node, whereby T represents just one tensor. The model provides stronger results, according in relation to a tensor structure, that are more instinctively more straightforward than computation and also has a reduced number of parameter sets. Furthermore, they characterize sentiment orientations over non-root nodes of lexical trees, helping them to capture sentiment transformations from phrases to statements more effectively.

Three alternative routes are available for the line of work. To begin, several studies have attempted to identify stronger tree composition operations. A variety of works, for example $v_p = f\,(W_1 v_l,\ W_2 v_r)$, literally use to compose the leaf nodes. The approach is much more straightforward, but it suffers from gradient explosion or diminish, making parameter learning incredibly difficult. Several experiments, inspired by the work of LSTM-RNN, propose the LSTM adaptation for recursive neural networks. [Tai et al. (2015) Tai, Socher & Manning] and [Chen et al. (2016) Chen, Zhu, Ling, Wei, Jiang & Inkpen] are two descriptive works which show the utility of LSTM across tree frameworks.

Second, multichannel compositions can improve a sentence simplification re- cursive neural network. The feasibility of such an upgrade is investigated by [Dong et al. (2014) Dong, Loy, He & Tang].

They use C homogeneous compositions to generate C hidden vector outputs, that are being used to reflect the source entity by integrating attention. They test the approach on basic recursive neural networks and find that it consistently outperforms some on a variety of benchmark datasets.

The third approach is to look at recursive neural networks using wider neural net-work system, close to what multilayer CNN researchers have done. In a nutshell, a recursive neural network is implemented over the input term embeddings as the first layer. When all of the hidden vectors in the output were formed, the same recursive neural network can be used again. [Irsoy & Cardie (2014)] also performed observational research on the process. The findings of the experiments show that a deeper recursive neural network will outperform a single-layer recursive neural network.

It's important to note that many studies use recursive neural networks to de- scribe sentences without using syntactic tree structures. Centered on raw sentence inputs, these findings suggest pseudo tree architectures. Furthermore, [Chen et al. (2015) Chen, Xu, Liu, Zeng & Zhao] use a simplified approach to automatically construct a tree structure for a sentence. For sentence-level emotion analysis, both works produce competitive results.

4. Sentiment Analysis at the Document Level

The aim of document-level sentiment classification is to determine a document's sentiment mark [Pang et al. (2002) Pang, Lee & Vaithyanathan]. The emotion marks could be two-dimensional, such as happy face and sad face [Pang et al. (2002) Pang, Lee & Vaithyanathan], or multi-dimensional, such as 15 stars on review pages [Pang et al. (2002) Pang, Lee & Vaithyanathan]. Current emotion classification methods in the literature can be divided into two categories: lexicon-based and corpus-based.

[Taboada et al. (2011) Taboada, Brooke, Tofiloski, Voll & Stede] provides an example of a lexicon-based system, which consists of three stages. If the POS tags of the phrases match the predefined patterns, they are extracted first. The emotion polarity of each derived expression is then calculated using point wise reciprocal knowledge (PMI), a statistical calculation of statistical dependency between two phrases. The PMI factor is obtained in Turney's work by pouring queries into a search query and looking at the number of hits. Finally, he calculates the sentiment polarity of a review by averaging the polarity of all phrases in it. To improve the efficiency of the lexicon-based approach, [Ding et al. (2008) Ding, Liu & Yu] use negation terms like "non", "never", and "cannot," as well as contrary words like "but." [Taboada et al. (2011) Taboada, Brooke, Tofiloski, Voll & Stede] incorporate intensifications including negation expressions of emotion lexemes that include angles and opinion advantages annotated.

The assumption behind designing a neural network solution is that function engineering is time-consuming. Instead, neural network methods can extract explanatory factors from input, reducing the need for robust feature extraction in learning algorithms.

[Bespalov et al. (2011) Bespalov, Bai, Qi & Shokoufandeh] embed each term as a vector and then use a perceptual convolutional network to extract the vectors for phrases. The phrase vectors are averaged to determine the text embedding. To discover the embeddings of claims and documents each text is represented by a dense vector that was already programmed to anticipate terms in the text. To predict the middle expression, the PV-DM model expands by mixing the document vector against ambient variables, the skip-gram system is created. They use sentence vectors to compose the text vector after modelling the embedding of sentences from terms. [Denil et al. (2014) Denil, Demiraj & De Freitas] use the same convolutional neural network for the sentence and text modelling components to measure the sentence vector with a convolutional neural network, then the text embedding with a bi-di rectional gated recurrent neural network.

Also, there are analyses that look at side details including user expectations or overall product quality to enhance document-level sentiment classification. [Ding et al. (2008) Ding, Liu & Yu] utilize an existing convolutional neural network to integrate user-sentiment accuracy and client consistency.

5. Sentiment Analysis on a Finer Scale

We present recent developments in fine-grained sentiment analysis using deep learning in this section. Fine-grained sentiment analysis, unlike sentence/document level sentiment classification, entails a variety of functions, the majority of which have distinct characteristics. As a result, these functions are modelled accordingly, taking into account their unique programmed environments. Among other fine-grained sentiment analysis subjects, we include opinion mining, personalized sentiment analysis, aspect-level sentiment analysis, stance identification, and sarcasm detection.

5.1. Opinion Mining

Opinion mining, which attempts to extract organized viewpoints from user-generated feedback, has become a hot topic in the NLP culture. The role usually consists of two fundamental sets of subtasks. First, we identify owners are examples of opinion bodies, aims, and expressions, and then we create relationships between them, such as the IS-ABOUT connection, that identifies the goal of a different perspectives, and the IS-FROM relation, which connects a personal experience to its speaker. Besides that, classifying emotion orientations is a critical role.

Opinion mining is a basic functional learning challenge that has indeed been thoroughly researched using standard mathematical models and discrete roles produced by humans. Using neural networks, we explain some representative studies of this activity in the following sections.

The initial stuff on neural network models aims to detect opinion entities, approaching the challenge as a sequence marking issue to identify opinion entity boundaries. For the mission, [Irsoy & Cardie (2014)] analyses the RNN structure. They use Elman-type RNNs to investigate the usefulness of bidirectional RNNs and the impact of RNN depth. Their findings suggest that bi-directional RNNs do well, with a three- layer bi-directional RNN achieving the best outcomes.

[Liu et al. (2019) Liu, He, Chen & Gao] also suggested a similar work. They investigate RNN alterations, such as Elman-type RNNs, Jordan-type RNNs, and LSTMs, in depth. They're still looking at bidirectionality.

5.2. Sentiment Analysis with a Purpose

[Dong & Potenza (2014)] suggest the first ever neural net model f or targeting-dependent emotion analysis. The model is based on [Dong et al. (2014) Dong, Loy, He & Tang]'s previous study, which we discussed in the emotion analysis at the sentence stage. Likewise, individuals generate recursive learning algorithms based on a binary contingent tree structure through using micro from the child nodes. The above work differs because although they transform the dependent tree based on the input target, having the target's headword the root of the resulting tree rather than the initial head word of the input sentence.

Automatic syntactic parsers generate input dependence parsing trees, which are heavily used in the above work. The trees can contain errors, resulting in an error propagation problem. Recent studies recommend performing selective sentiment analysis with only raw sentence inputs to prevent the issue. To extract a range of neural features for the mission, [Vo & Zhang (2015)] use a variety of pooling techniques. The neural features that arise are concatenated to forecast emotion polarity.

Several recent studies have looked at the utility of RNN for the job, which has shown positive results in other sentiment analysis tasks. [Yang et al. (2017) Yang, Zhang & Dong] consider using gated RNN to improve sentential word representation. The resulting representations will grab context- sensitive information thanks to the use of RNN. [Dong et al. (2016) Dong, Loy & Tang] use LSTM-RNN as a single simple neural layer to encode sequential input terms. In terms of selective emotion analysis, both works have obtained good results.

5.3. Sentiment Analysis at the Aspect Level

The intent of elemental sentiment analysis is to categories the orientations of sentiment in a sentence. A feature is a property of a goal that allows humans to articulate their feelings about it. Typically, the job entails analyzing consumer feedback for a specific product, such as a restaurant, appliances, or a film. Products may have a variety of features. For example, the setting, price, and service are all facets of a hotel, and users typically leave reviews to share their opinions about each. Aspects should be enumerated when the product is given, unlike focused sentiment analysis, and the aspect cannot be articulated consistently in one evaluation in some cases.

Since the objective is initially modelled as a statement classifier challenge, we can use the same approach as we did for sentence-level emotion classification, excluding the fact also that classes were distinct. As every factor might just have three orientations of opinion: optimistic, destructive, and unfavorable, aspect-level classification technique is usually a 3N classification task, assuming that a substance has N predefined aspects. For the mission, [Lakkaraju et al. (2014) Lakkaraju, Socher & Manning] suggest a matrices formulation built on an iterative neural network paradigm, close to that proposed by [Socher et al. (2012) Socher, Huval, Manning & Ng] for sentence- level emotion classification.

In real-life situations, a single feature of a product may take on many different forms. Using a laptop as an example, the screen can be expressed in terms of display, resolution, and appearance, all of which are closely linked to screen. The findings that the sentiment analysis at the addition to the interest seems to be more helpful for more use if we can group related aspect phrases into one aspect. The first neural network model for aspect word classification is proposed by [Lakkaraju et al. (2014) Lakkaraju, Socher & Manning]. They use basic multilayer feed-forward neural networks to learn representations of aspect phrases, using attention structure to identify neural characteristics. The model parameters are learned using automated testing examples and remote control. [Xiao et al. (2017) Xiao, Ye, He, Zhang, Wu & Chua] use an unsupervised auto- encoder method for retrieval of aspects, that uses an attention process to learn the size of aspect terms automatically.

5.4. Stance Detection for the Textual Dataset

The aim of stance detection is to identify a sentence's attitude toward a specific subject. In several instances, the task's object is identified into one source, and indeed the statement to have been categorized on the other. It's possible that the input sentences don't have any clear connections to the given subject. As a result, detecting posture is exceedingly challenging.

For each subject, the early work trains independent classifiers. As a result, the challenge is viewed as a straightforward three-way classification dilemma.

5.5. Sarcasm Identification

In this segment, we'll look at a unique language phenomenon called sarcasm or irony, which has a lot to do with sentiment analysis. This phenomenon alters the literal sense of a sentence and has a significant impact on the emotion conveyed by the sentence [Pandey et al. (2021) Pandey, Kumar, Singh & Tripathi]. A simple dataset consists of sarcastic and non-sarcastic tweet posted by the different users can be seen from Figure 1.

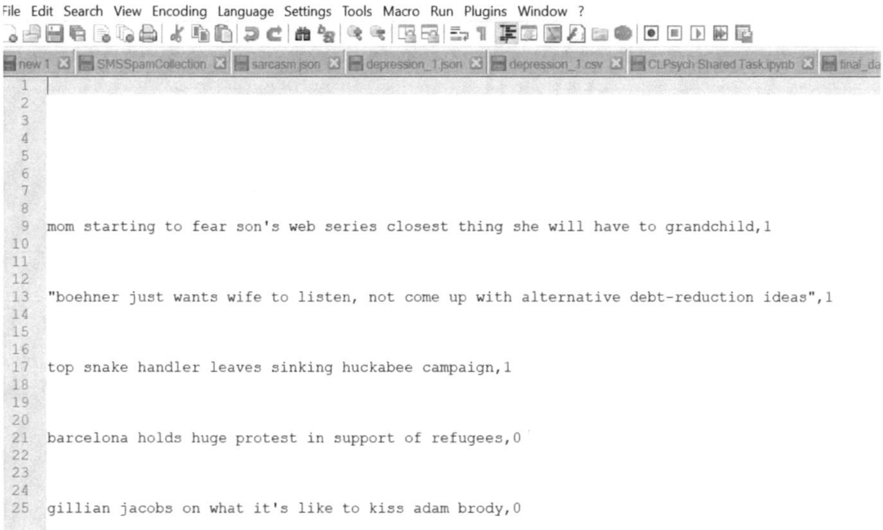

Figure 1. Instance of a tweet dataset consisting of sarcastic and non-sarcastic text.

Sarcasm identification is typically modelled into a two-class problem, which is equivalent to sentence-level sentiment analysis. The only distinction between the two activities is in their objectives. [Ghosh & Veale (2016)] investigate a variety of neural network models for the mission, including CNN, LSTM, and deep feed-forward neural networks, in depth. They introduce a variety of neural networks and empirically test their efficacy. The findings of the experiments demonstrate that combining these neural networks produces the best results. A two-layer CNN, a two-layer LSTM, and one feed-forward layer make up the final model.

Author-based knowledge is one kind of useful function for detecting sarcasm in social media like Twitter. For Twitter sarcasm detection, [He et al. (2016) He, Zhang, Ren & Sun] suggest a contextually relevant neural model. They extract a selection of key terms from the tweet authors' previous posts and use these words to describe the tweet speaker. The two aspects of their proposed neural network model are a gated RNN for representing sentences and a basic pooling neural network for representing tweet writers.

CONCLUSION

We provide a quick recap of the latest progress of neural network methods in sentiment examination in this chapter. To learn sentiment-specific word embeddings, we first explain how to integrate sentiment knowledge from texts. Then we go through sentence and document sentiment classification, which all involve semantic text composition. Then we'll show you how to build neural network models for fine-grained tasks.

Despite the fact that deep learning methods have shown impressive results on sentiment analysis tasks in recent years, there are several possible ways to develop this field further. Sentiment analysis that makes sense is the first step towards the sentiment extraction. Deep learning models are currently available that are reliable but inexplicable. Using cognitive science information, common sense knowledge, or derived knowledge from a text corpus may be a way to develop this field. Learning a stable model for a new domain is the second path. The volume and consistency of training data determine the success of a deep learning algorithm. As a result, learning a robust sentiment analyzer for a domain with little/no annotated corpus is difficult but crucial for real-world applications. The third path is to learn how to comprehend feeling. The majority of current research focuses on expressions of opinion, goals, and owner. New characteristics, such as viewpoint causes and stances, have recently been proposed to help explain sentiment. The advancement in this field necessitates the use of strong models and massive corpora. The next approach is fine-grained emotion analysis, that have recently gained popularity. A greater training corpus is needed to improve this region [Pang & Lee (2006)].

REFERENCES

[Almeida & Xexéo (2019)] Almeida, F. & Xexéo, G. (2019). Word embeddings: A survey. arXiv preprint *arXiv:1901. 09069*,

[Alsaeedi & Khan (2019)] Alsaeedi, A. & Khan, M. Z. (2019). A study on sentiment analysis techniques of twitter data. *International Journal of Advanced Computer Science and Applications*, *10*, 361–374.

[Alswaidan & Menai (2020)] Alswaidan, N. & Menai, M. E. B. (2020). A survey of state-of-the-art approaches for emotion recognition in text. *Knowledge and Information Systems*, (pp. 1–51).

[Arora & Kansal (2019)] Arora, M. & Kansal, V. (2019). Character level embedding with deep convolutional neural network for text normalization of unstructured data for twitter sentiment analysis. *Social Network Analysis and Mining*, *9*, 1– 14.

[Batrinca & Treleaven (2015)] Batrinca, B. & Treleaven, P. C. (2015). Social media analytics: a survey of techniques, tools and platforms. *Ai & Society*, *30*, 89–116.

[Berka (2020)] Berka, P. (2020). Sentiment analysis using rule-based and case-based reasoning. *Journal of Intelligent Information Systems*, (pp. 1–16).

[Bespalov et al. (2011) Bespalov, Bai, Qi & Shokoufandeh] Bespalov, D., Bai, B., Qi, Y. & Shokoufandeh, A. (2011). Sentiment classification based on supervised latent n-gram analysis. In *Proceedings of the 20th ACM international conference on Information and knowledge management*, (pp. 375–382).

[Chen et al. (2018) Chen, Ling & Zhu] Chen, Q., Ling, Z. H. & Zhu, X. (2018). Enhancing sentence embedding with generalized pooling. arXiv preprint *arXiv:1806.09828*.

[Chen et al. (2016) Chen, Zhu, Ling, Wei, Jiang & Inkpen] Chen, Q., Zhu, X., Ling, Z., Wei, S., Jiang, H. & Inkpen, D. (2016). Enhanced lstm for natural language inference. arXiv preprint *arXiv:1609.06038*.

[Chen et al. (2015) Chen, Xu, Liu, Zeng & Zhao] Chen, Y., Xu, L., Liu, K., Zeng, D. & Zhao, J. (2015). Event extraction via dynamic multi-pooling convolutional neural networks. In *Proceedings of the 53rd Annual Meeting of the Association for Computational Linguistics and the 7th International Joint Conference on Natural Language Processing (Volume 1: Long Papers)*, (pp. 167–176).

[Das et al. (2017) Das, Ganguly & Garain] Das, A., Ganguly, D. & Garain, U. (2017). Named entity recognition with word embeddings and wikipedia categories for a low-resource language. *ACM Transactions on Asian and Low-Resource Language Information Processing (TALLIP)*, *16*, 1–19.

[Denil et al. (2014) Denil, Demiraj & De Freitas] Denil, M., Demiraj, A. & De Freitas, N. (2014). Extraction of salient sentences from labelled documents. arXiv preprint *arXiv:1412.6815*.

[Ding et al. (2008) Ding, Liu & Yu] Ding, X., Liu, B. & Yu, P. S. (2008). A holistic lexicon-based approach to opinion mining. In *Proceedings of the 2008 interna- tional conference on web search and data mining* (pp. 231–240).

[Dong et al. (2014) Dong, Loy, He & Tang] Dong, C., Loy, C. C., He, K. & Tang, X. (2014). Learning a deep convolutional network for image super-resolution. In *European conference on computer vision*, (pp. 184–199). Springer.

[Dong et al. (2016) Dong, Loy & Tang] Dong, C., Loy, C. C. & Tang, X. (2016). Accelerating the super-resolution convolutional neural network. In *European con- ference on computer vision*, (pp. 391–407). Springer.

[Dong & Potenza (2014)] Dong, G. & Potenza, M. N. (2014). A cognitive-behavioral model of internet gaming disorder: Theoretical underpinnings and clinical implications. *Journal of psychiatric research*, *58*, 7–11.

[Ghosh & Veale (2016)] Ghosh, A. & Veale, T. (2016). Fracking sarcasm using neural network. In *Proceedings of the 7th workshop on computational approaches to subjectivity, sentiment and social media analysis* (pp. 161–169).

[Gu et al. (2018) Gu, Wang, Kuen, Ma, Shahroudy, Shuai, Liu, Wang, Wang, Cai et al.] Gu, J., Wang, Z., Kuen, J., Ma, L., Shahroudy, A., Shuai, B., Liu, T., Wang, X., Wang, G., Cai, J. et al. (2018). Recent advances in convolutional neural networks. *Pattern Recognition*, *77*, 354–377.

[He et al. (2016) He, Zhang, Ren & Sun] He, K., Zhang, X., Ren, S. & Sun, J. (2016). Identity mappings in deep residual networks. In *European conference on computer vision*, (pp. 630–645). Springer.

[Irsoy & Cardie (2014)] Irsoy, O. & Cardie, C. (2014). Opinion mining with deep recurrent neural networks. In *Proceedings of the 2014 conference on empirical methods in natural language processing (EMNLP)*, (pp. 720–728).

[Jacovi et al. (2018) Jacovi, Shalom & Goldberg] Jacovi, A., Shalom, O. S. & Gold- berg, Y. (2018). Understanding convolutional neural networks for text classification. arXiv preprint *arXiv:1809.08037*.

[Jianqiang et al. (2018) Jianqiang, Xiaolin & Xuejun] Jianqiang, Z., Xiaolin, G. & Xuejun, Z. (2018). Deep convolution neural networks for twitter sentiment analysis. *IEEE Access*, *6*, 23253–23260.

[Kalchbrenner et al. (2017) Kalchbrenner, Grefenstette & Blunsom] Kalchbrenner, N., Grefenstette, E. & Blunsom, P. (2017). A convolutional neural network for modelling sentences (2014). arXiv preprint *arXiv:1404.2188*.

[Kim (2019)] Kim, Y. (2019). Convolutional neural networks for sentence classification. arxiv 2014. arXiv preprint *arXiv:1408.5882*.

[Lakkaraju et al. (2014) Lakkaraju, Socher & Manning] Lakkaraju, H., Socher, R. & Manning, C. (2014). Aspect specific sentiment analysis using hierarchical deep learning. In *NIPS Workshop on deep learning and representation learning*, (pp. 1–9).

[Li et al. (2018) Li, Liu, Chen & Rudin] Li, O., Liu, H., Chen, C. & Rudin, C. (2018). Deep learning for case-based reasoning through prototypes: A neural network that explains its predictions. In *Proceedings of the AAAI Conference on Artificial Intelligence*. volume *32*.

[Liu et al. (2019) Liu, He, Chen & Gao] Liu, X., He, P., Chen, W. & Gao, J. (2019). Multi-task deep neural networks for natural language understanding. arXiv preprint *arXiv:1901.11504*.

[Manning & Schutze (1999)] Manning, C. & Schutze, H. (1999). *Foundations of statistical natural language processing*. MIT press.

[Pandey et al. (2021) Pandey, Kumar, Singh & Tripathi] Pandey, R., Kumar, A., Singh, J. P. & Tripathi, S. (2021). Hybrid attention-based long short-term memory network for sarcasm identification. *Applied Soft Computing*, (p. 107348).

[Pang & Lee (2006)] Pang, B. & Lee, L. (2006). Opinion mining and sentiment analysis. foundations and trend. *Information Retrieval*, *2*.

[Pang et al. (2002) Pang, Lee & Vaithyanathan] Pang, B., Lee, L. & Vaithyanathan, S. (2002). Thumbs up? sentiment classification using machine learning techniques. *arXiv preprint cs/0205070*.

[Ruas et al. (2020) Ruas, Ferreira, Grosky, de Franc¸a & de Medeiros] Ruas, T., Fer- reira, C. H. P., Grosky, W., de Franc¸a, F. O. & de Medeiros, D. M. R. (2020). Enhanced word embeddings using multi-semantic representation through lexical chains. *Information Sciences*, *532*, 16–32.

[Schmidt et al. (2019) Schmidt, Marques, Botti & Marques] Schmidt, J., Marques, M. R., Botti, S. & Marques, M. A. (2019). Recent advances and applications of machine learning in solid-state materials science. *npj Computational Materials*, *5*, 1–36.

[Socher et al. (2012) Socher, Huval, Manning & Ng] Socher, R., Huval, B., Manning, C. D. & Ng, A. Y. (2012). Semantic compositionality through recursive matrix- vector spaces. In *Proceedings of the 2012 joint conference on empirical methods in natural language processing and computational natural language learning*, (pp. 1201–1211).

[Socher et al. (2013) Socher, Perelygin, Wu, Manning, Ng & Potts] Socher, R., Pere- lygin, A., Wu, J., Manning, C. D., Ng, A. Y. &

Potts, C. (2013). Recursive deep models for semantic compositionality over a sentiment treebank. In *Proceedings of the 2013 conference on empirical methods in natural language processing*, (pp. 1631–1642).

[Taboada et al. (2011) Taboada, Brooke, Tofiloski, Voll & Stede] Taboada, M., Brooke, J., Tofiloski, M., Voll, K. & Stede, M. (2011). Lexicon-based methods for sentiment analysis. *Computational linguistics*, *37*, 267–307.

[Tai et al. (2015) Tai, Socher & Manning] Tai, K. S., Socher, R. & Manning, C. D. (2015). Improved semantic representations from tree-structured long short-term memory networks. arXiv preprint *arXiv:1503.00075*.

[Vo & Zhang (2015)] Vo, D. T. & Zhang, Y. (2015). Target-dependent twitter sentiment classification with rich automatic features. In *Twenty-fourth international joint conference on artificial intelligence*.

[Wang et al. (2015) Wang, Liu, Sun, Wang & Wang] Wang, X., Liu, Y., Sun, C. J., Wang, B. & Wang, X. (2015). Predicting polarities of tweets by composing word embeddings with long short-term memory. In *Proceedings of the 53rd Annual Meeting of the Association for Computational Linguistics and the 7th In- ternational Joint Conference on Natural Language Processing (Volume 1: Long Papers)*, (pp. 1343–1353).

[Xiao et al. (2017) Xiao, Ye, He, Zhang, Wu & Chua] Xiao, J., Ye, H., He, X., Zhang, H., Wu, F. & Chua, T. S. (2017). Attentional factorization machines: Learning the weight of feature interactions via attention networks. arXiv preprint *arXiv:1708.04617*.

[Yang et al. (2017) Yang, Zhang & Dong] Yang, J., Zhang, Y. & Dong, F. (2017). Neural word segmentation with rich pretraining. arXiv preprint *arXiv:1704.08960*.

[Yin & Schtze (2015)] Yin, W. & Schtze, H. (2015). *Multichannel variable-size convolution for sentence classification*., (pp. 204–214).

[Young et al. (2018) Young, Hazarika, Poria & Cambria] Young, T., Hazarika, D., Poria, S. & Cambria, E. (2018). Recent trends in deep learning based natural language processing. *IEEE Computational intelligence magazine*, *13*, 55–75.

[Zador (2019)] Zador, A. M. (2019). A critique of pure learning and what artificial neural networks can learn from animal brains. *Nature communications*, *10*, 1–7.

[Zhang et al. (2018) Zhang, Wang & Liu] Zhang, L., Wang, S. & Liu, B. (2018). Deep learning for sentiment analysis: A survey. *Wiley Interdisciplinary Reviews: Data Mining and Knowledge Discovery*, *8*, e1253.

In: Deep Learning and its Applications ISBN: 978-1-68507-185-1
Editor: Arvind Kumar Tiwari

Chapter 8

DEEP LEARNING TECHNIQUES IN PROTEIN-PROTEIN INTERACTION

***Bachcha Lal Pal*[1], *Akshay Deepak*[1] *and Arvind Kumar Tiwari*[2]**
[1]CSE Department, NIT Patna, Patna, Bihar, India
[2]CSE Department, KNIT Sultanpur, Sultanpur, Uttar Pradesh, India

ABSTRACT

Protein-Protein Interactions are essential in all organisms cellular functioning and biological processes. The majority of genes and proteins realize the resulting phenotype function as a set of interactions. protein-protein interactions are an important molecular process in cells and play a key role in their biochemical function. Whereas high-throughput experimental techniques have advanced, allowing researchers to detect large quantities of protein-protein interaction, they are not without drawbacks, deep learning techniques have been widely used to predict protein-protein interaction, thus allowing experimental researchers to study protein-protein interaction networks. This paper provides a brief study of methods used in Protein-Protein Interactions, their limitations, and their applications of Protein-Protein Interactions.

Keywords: protein-protein interaction, deep learning technique

1. INTRODUCTION

Protein-Protein Interactions (PPIs) are fast becoming one of system biology's most important goals. Non-covalent interactions between residue side chains are used in protein folding, protein assembly, and PPI [1]. The proteins from various interactions and associations as a result of these encounters. PPIs can be characterized in various ways based on their structural and functional differences. Different types of protein interactions accomplish the majority of critical molecular functions in cells. As a function, determining the protein-protein interactions of organisms is one of the primary goals of functional proteomics. Proteomics research using PPIs as the major study content is now frequently used in the discovery of medical drug targets. Promoting the development of the biomedical industry is highly significant. Protein interactions are governed via interfaces, which are amino acid connections between various proteins. Finding protein interfaces necessitates identifying all amino acid pairings complex and challenging [2]. To prediction the protein-protein interaction, computational methods have been proposed [3-6]. These strategies are based on handcrafted highlights sorted from a few spaces, which are used at that point to estimate interaction using traditional machine learning procedures. To accomplish particular actions, a protein interacts with other proteins. PPI occurs at nearly all levels of cellular function. The discovery of protein interactions gives a comprehensive picture of cellular functions and biological processes. Because many biological processes require PPIs, pinpointing the collection of interacting proteins in an organism is critical for understanding the molecular mechanisms behind certain biological functions and assigning functions [7-9]. Predicting protein interactions is also a crucial step in constructing PPI networks for humans and other species. The discovery of potential viral-host protein interactions could aid in a better understanding of infectious mechanisms, leading to the development of new drugs and treatment options. PPI abnormalities have been linked to various neurological illnesses, including Creutzfeldt-Jakob disease and Alzheimer's disease [10-12]. As a result, developing accurate and trustworthy approaches for finding PPIs has far-reaching implications in the diversity of protein research fields. Deep learning algorithms effectively determine high-dimensional and non-linear characteristics from various sources, including protein sequences, real-world applications, pattern recognition, and image identification [13, 14]. It is necessary to incorporate several modalities (structural, First, and Second-order similarity) from the protein network, to execute deep learning and achieve high performance in the PPI analysis [15].

PPIs play a key role in various biological processes, including enzyme-substrate interactions, hormone-receptor binding, and immunological functions such as antigen recognition and presentation. To name a few, disclosing that an altered PPI can cause disease onset. Analyzing the molecular details of PPI will add to our understanding, but it will also be helpful in current medical biochemistry fields such as molecular medicine and other research areas.

The remaining chapter is organized as follows. Section 2 elaborates the definition of protein. After that definition of protein-protein interaction and types of protein-protein interaction are discussed in sections 3 and 4, respectively. In section 5, we provide a brief discussion on the methodologies used in PPIs. Furthermore, the challenges of PPIs and applications of PPIs are illustrated in sections 6 and 7. And finally, section 8 concludes the chapter.

2. PROTEIN

A protein is made of one or more amino acid chains. Proteins, often known as macromolecules, are massive biomolecules. Proteins are made up of twenty different amino acids in general. Proteins can serve as antibodies, contractile proteins, structural proteins, enzymes, transport proteins, and storage proteins, among other things.

Amino acids are divided into twenty Proteinergic Amino acids based on the physicochemical characteristics of their side chains.

In the protein chain, amino acids form a highly stable covalent bond called a peptide bond (-CO-NH-) between the carboxylic group of the previous residue and the amino group of the next residue. Different amino acid side chains are essential for PPI because they actively participate in it.

3. Protein-Protein Interaction

Proteins in biological systems are also commonly found in oligomeric forms than in monomeric or isolated forms. Oligomeric proteins are protein pairs made up of two or more proteins. While a dimer is a preferred term for a two-protein oligomer. Protomer refers to every oligomer's essential protein chain. As a response, conserved amino acid residues were found at interacting surfaces (similar to a hydrophobic protein core interacting with polar surfaces). PPI, a basic cellular physical mechanism, is involved in a variety of cellular physiology and metabolisms. Channels and receptor functions: On the cell surface, receptors such as tyrosine kinase phosphates alpha behave primarily as a weak homodimer, with one monomer intruding on the active site of the other. This prevents molecules from being transferred unnecessarily.

Combinatorial specificity is controlled. The specific activity of enzymes such as protein tyrosine phosphatases is regulated in a combinatorial manner by substrate target domains and enzyme catalytic domains. During signal transmission, signaling protein molecules exhibit the same form of combinatorial specificity.

4. Types of Protein-Protein Interaction

The Protein-Protein Interaction can be divided into several groups as presented in Figure 1.

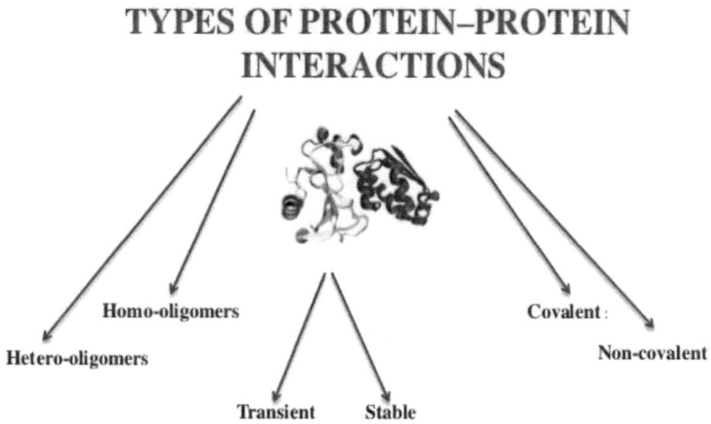

Figure 1. Types of Protein-Protein Interaction.

Homo-Oligomers

Homo-oligomers are macromolecular complexes made up entirely of the same kind of protein subunit. The formation of non-covalent interactions in the protein's quaternary structure guides the assembly of protein subunits.

Hetero-Oligomers

Hetero-oligomers are needed for a variety of cellular functions. Contact between heterologous proteins is much more important during cell signaling events, and such interactions are only possible because of structural domains within the proteins.

Stable

Proteins that interact for a long time as subunits of permanent complexes to carry out functional roles are called stable interactions.

Transient

Many aspects of cellular function depend on transient interactions, which are protein interactions that shape and break quickly.

Covalent

Disulphide bonds or electron sharing form covalent interactions, which have the strongest connection.

Non-Covalent

When weaker bonds like hydrogen bonds, ionic interactions, Vander Waals powers, or hydrophobic bonds, are combined during transient interactions, non-covalent bonds are formed.

5. Methodologies Used in Protein-Protein Interaction

The literature has been studied and described for computational-based PPI investigations, machine, and deep learning methodologies.

5.1. Deep Learning

Deep learning, a common source of machine learning that builds on the original neural networks popularised in the 1980s, is a popular machine learning type. As compared to the first neural network, there is minimal variation in how a deep neural network is measured. Deep neural networks have always been possible to build and quantify. A deep neural network is a neural network with a large number of layers. We have always been able to create/calculate deep neural networks, but we have never had an excellent way to train them. Deep learning is an effective tool for training deep neural networks.

Deep Learning is a subset of artificial neural networks used in deep learning to conduct complex computations on large quantities of data. It's a form of machine learning that's focused on the human body's structure and function.

Machines are trained using deep learning algorithms that learn from examples. Deep learning is widely used in industries such as medical management, e-commerce, advertisement, etc.

Although Feed Forward Neural Networks (FFNN) or similar elementary cells are frequently employed as the basic building blocks of a Deep Learning system, they are integrated into deep stacks utilizing different connectivity patterns. Deep Learning models can be customized for any kind of data according to this architectural flexibility. Back-propagation [16] can be adopted in the training process of deep learning models for example As a result, powerful internal representations of the data learned for a task are produced. Deep learning models can be trained on examples using back-propagation, resulting in powerful internal representations of the data being learned for a task. This automated feature learning eliminates the requirement for manual, feature engineering, which is a time-consuming and potentially error-prone process that requires the domain specialist for approaches of machine learning. Deep Learning models can easily have a large number of internal parameters and are, therefore: data-hungry; until now, the most promising Deep Learning implementations have in fields with a large number of examples [17].

5.2. Approaches of Deep Learning

Supervised learning, unsupervised learning, reinforcement learning, and hybrid learning are all effective with deep neural networks.

Supervised Learning

In supervised learning, the input variables represented as X are mapped to out-put variables represented as Y through using a set of rules to study the mapping feature f.

$$Y = f(X)$$

The learning algorithm aims to predict the output (Y) for new input by approximating the mapping function (X The error from the predictions made during training can be used to correct the output. Learning can be stopped once all inputs have been trained to produce the desired output. [18, 19]. Support Vector Machines (SVMs) are used to classify data [20].

Unsupervised Learning

In this, there is no corresponding output variable from the mapping of the input variable. Unsupervised learning is used to solve clustering and association problems. In clustering challenges, unsupervised learning approaches are utilized [21], for example, K-means clustering, etc.

Hybrid Learning

The hybrid learning architectures combine generative (unsupervised) and discriminative (supervised) learning components. A hybrid deep neural network can be created by combining different architectures. They are utilized to recognize human actions by action bank features, and they are projected to generate significantly improved results [22].

Reinforcement Learning

To train the algorithm, reinforcement learning employs a reward and punishment mechanism. The algorithm or agent learns from its surroundings in this case. When the agent performs correctly, he is rewarded, and when he performs incorrectly, he is penalized. Consider the case of a self-driving car, where the agent is rewarded for arriving on time and penalized for going off-road. The agent wants to enhance the benefit while minimizing the cost. In reinforcement learning, the algorithm is not informed how to learn; instead, it solves the problem on its own [24].

5.3. Deep Learning Technique

The following section discusses some of the powerful methods that can be applied to deep learning algorithms to reduce training time and optimise the model. The summarizes the advantages and disadvantages of each technique in Table 1.

Stochastic Gradient Descent

In the gradient descent algorithms, using the convex function ensures that an optimal minimum is found without becoming trapped in a local minimum. It may arrive at the optimum value in a variety of ways, depending on the functions parameters and the learning rate or step size [25].

Batch Normalization

Batch normalisation minimises covariate shift, allowing deep neural networks to run faster. When the weights are adjusted throughout the training, it normalises the inputs to a layer for each mini-batch. Normalization decreases training epochs and learning. To improve the stability of a neural network, the output from the preceding activation layer can be normalised [26].

Back Propagation

The backpropagation can be used to find the employing gradient of the function for each iteration when using a gradient-based technique to solve an optimization problem [27].

Max-Pooling

Alter is pre-denied in max-pooling, and this letter is then applied across the non-overlapping sub-regions of the input, with the output being the maximum of the values included in the window. Max-pooling can minimize dimensionality as well as the computing cost of learning certain parameters [28].

Dropout

Deep neural networks, the drop-out strategy can be used to tackle the overtaking problem. During training, this strategy is used by dropping units and their connections at random [9]. Dropout is a useful regularization technique for reducing overfitting and improving generalization error. Dropout enhances performance on supervised learning tasks in computational biology, computer vision, document classification, and speech recognition [29].

Transfer Learning

In transfer learning, a model trained on one task is used for another task that is related to all of them. The knowledge gained from addressing a specific challenge can be transmitted to another network that will be trained on a similar challenge. While working on the second problem, this provides for faster development and better results [30].

Skip-Gram

Skip-gram can be used to model word embedding methods. When two vocabulary terms have a similar context in the skip-gram model, they are considered identical. For example, the sentences "dogs are mammals" and "cats are mammals" are both true and have the same meaning as "are mammals." A skip-gram is made by taking a context window with n phrases, training a neural network by skipping one of these words, and then using the model to predict the skipped term [31].

Neural Network

Neural Networks are a collection of structured methods to identify patterns and closely imitate the human brain. They use computer perception to interpret sensory data, labeling, or raw clustering data. They can find hidden numerical patterns in vectors, which must then be transformed into all real-world data (text, sound, image, or time series). Artificial neural networks (ANNs) are made up of several highly interconnected computing elements (neurons) that collaborate to solve a problem.

Table 1. Deep learning techniques comparison

Deep Learning Technique	Description	Advantage	Disadvantage
Stochastic Gradient Descent	To Find optimal minimal optimization Problems	Avoids entanglement in local minimums.	Convergence time is longer, and it is more computationally costly.
Batch Normalization	Batch normalization of input to a layer	Reduces covariant shift and improves network stability. Network trains are more effective.	During preparation, there is a computational overhead.
Max-Pooling	The maximum filter is applied.	Reduce the dimensions and computation cost.	Considers just the essential factor, which in some situations can result in an undesirable outcome.
Back Propagation	Used in an issue of optimization	To compute the gradient	Data that is sensitive to noise
Dropout	During instruction, dropout units/connection	Avoid overfitting	Increases the number of iterations required to reach convergence.
Transfer Learning	The first model's knowledge is applied to the second problem.	Improves efficiency and allows for rapid improvement in the preparation of the second problem.	It just works on identical issues.
Skip-gram	Algorithms for word embedding	Can work with any unprocessed text It necessitates less memory.	Softmax is a computationally costly function with a long training time.

A Neural Network structure of the human brain consists of artificial neurons also, known, as nodes as shown in Figure 2. These nodes are placed in three levels close to each other.

- Input Layer
- Hidden Layers
- Output Layer

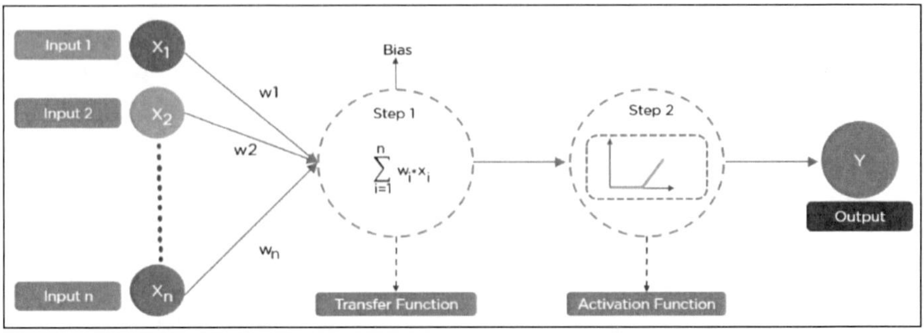

Figure 2. Neural Network.

The data provides information in the form of input to each node. The nodes take numerous random-weighted inputs, calculate them, and add a bias. Finally, to determine which neurons to fire, nonlinear functions, also known as activation functions, are used.

Convolutional Neural Networks

The architecture of the Convolutional Neural Network (CNN) [32] is built to process information that is structured with daily spatial information Dependency (For example, series tokens or pixels in a picture sequence). A CNN layer exploits this regularity by employing the same set of local convolutional filters across all data points, resulting in two advantages. It is translation invariant and avoids the overfitting problem by providing a restricted number of weights to alter the input layer and the dimensionality of the following layer.

A CNN module is typically made up of multiple consequential CNN layers, allowing the nodes in subsequent layers to have larger receptive fields. It's also possible to exchange it for more complex features.

Recurrent Neural Network

Recurrent neural networks (RNNs) are the type of neural networks that can be used to model sequence data. RNNs, which are derived from feed-forward networks, behave similarly to human brains. Simply put, recurrent neural networks can predict sequential data in a way that other algorithms can't. With the continued advancement of artificial neural network research [33], many challenging practical problems in the fields of biology, medicine, intelligent robots, pattern recognition, automatic control, and economics have been successfully solved. A neural network that models sequence data is the recurrent neuron network [34]. RNN has shown exceptional success in NLP, image, and speech recognition in recent years. The framework between levels in a traditional neural network model is completely linked, but the neurons within each layer are isolated for certain issues, this form of the neural network is ineffective.

Long Short Term Memory Networks: (LSTMs)

It is a form of recurrent neural network that extends the memory capacity. As a result, it is well adapted to learning from significant interactions separated by long periods.

The unit of an LSTM is used to construct an LSTM network, which is a layer of an RNN. RNNs can remember inputs for a long time compared to LSTMs. Because of LSTMs store information in memory similar to that of a machine. The LSTM's memory can be read, written, and erased. This memory can be thought of as a gated cell, with gated indicating that the cell decides whether to store or erase data (i.e., whether to open the gates) based on its value to the data. The algorithm also learns weights, which are used to allocate importance. Simply defined, it learns over time what information is relevant and what information is not.

There are three gates in an LSTM: input, forget, and output. These gates decide whether the new input should be allowed (input gate), whether it should be deleted because it isn't necessary (forget gate), or whether it should have an effect on the output at the current time step (impact gate) (output gate). The three gates of an RNN are depicted in the diagram below:

An LSTM's gate is analog in the form of a sigmoid, which means they range from zero to one. Since they are analog, they can perform back-propagation.

LSTM solves vanishing gradients' problem since it holds the gradients steep enough, resulting in a short training period and high accuracy.

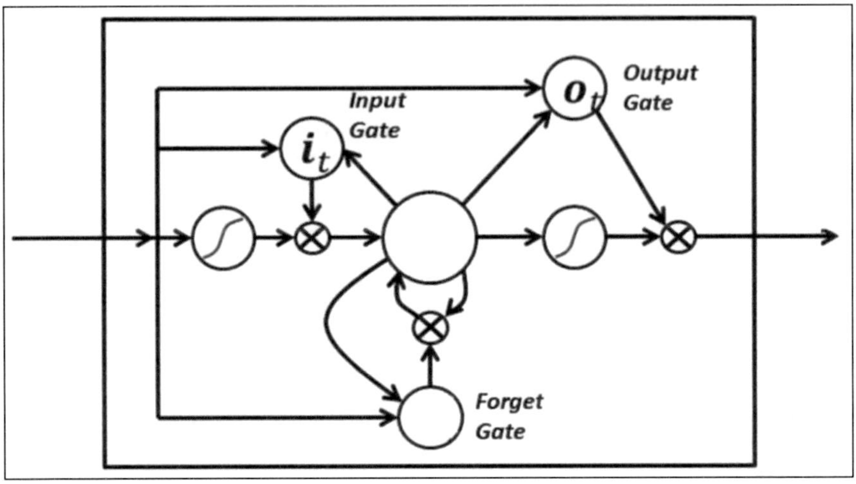

Figure 3. Long short term memory.

RNNs will use past knowledge if the gap between relevant information and expected location is small, but as the time interval grows, regular RNNs will be unable to learn long-distance information. LSTM neural network [18, 19] has been argued that long-term reliance can be taught as a solution to this issue. The main difference between the LSTM and other networks is the use of sophisticated memory blocks rather than general neurons. Three multiplicative "gate" units (input, forget, and output gates), as well as several memory cells, make up the memory block (one or more). The information flow is regulated by the gate unit, whereas the information flow is controlled by the memory cell. Information from the past should be preserved. The gate removes or restores data to the state of the cell by controlling the flow of information. The information input and output, which is more precise, Flow is in charge of the input and output gates, respectively. The forgotten gate controls the amount of data from the previous unit that is stored in the current unit [38-40].

6. Challenges and Issues

The computational study of protein sequences in general and PPI predictive faces several challenges. First, protein chains are often lengthy, making experimental procedures difficult, time-consuming, and expensive to characterize. Second, for the training and evaluation of prediction techniques, extensive, comprehensive, and accurate benchmark datasets are necessary. Third, because computational PPI approaches rely on experimentally gathered data, any error in the experimental data will impact the computational PPI predictions. Protein representation is one of the problems of protein prediction algorithms. To create their classification models, protein prediction algorithms differ in protein representation and feature extraction. The sequences and the discrete model are the two types of models commonly used to represent protein samples. The simplest discrete model is AA composition, which is then normalized occurrence frequencies of the twenty native amino acids in a protein. Physiochemical characteristics of AA are approached by some. The pairwise similarity is used in other methods. Templates are used in certain techniques, while others rely on statistical machines and deep learning.

7. Application

The development of macromolecular structures and enzymatic complexes is dependent on protein-protein interactions. PPIs have emerged as exciting targets for rational drug design in recent years, owing to their excellent specificity, allowing researchers to target specific disease-related pathways. Large-scale screening procedures and high-throughput approaches like the yeast two-hybrid system for studying individual PPIs are two types of experimental methodologies for exposing the molecular recognition processes of different PPIs. Due to a variety of physicochemical issues, including transient dynamics and posttranslational modification (PTM). In order to increase PPI coverage and filter out false positives using confidence scores of protein interactions, in silico methods to efficiently locate PPIs and PPI sites are required. A large number of in silico prediction approaches for distinguishing PPIs from non-PPIs or PPIs sites have been proposed. The classification of prediction methods and their possible applications in rational drug design, hotspot prediction, and docking is based on sequence, structure, homology, domains, functional similarity, gene expression, and network topology. The pharmaceutical industry is still wary of using PPIs in therapeutic development. The enzymatic activity of PPIs is challenging to assess. With the availability of in silico structures matching to various states, the structural features of the interactions can be revealed. *In vitro* or in silico biophysical approaches that can be targeted, small molecule modulators are employed for drug ability and hotspot analysis. PPI sites and hotspot residues that have been empirically verified can be used to build small compounds that govern the effects of therapeutic and drug-able PPIs. Arrange portrayal and inspection provides a framework level understanding of medicinal activity and infection, multiple natures and provides a framework level understanding of infection multifarious nature [41].

CONCLUSION

Several preset approaches demonstrate how protein structure and PPIs are coordinated at multiple levels. These methods allow us to understand not only how a pathogenic protein interacts with its host on an atomic level, but also how such collaborations function in larger cell architecture. Combining correct configurations of positive and negative preparation set machine deep learning algorithms is used to forecast high confirmation linkages. This chapter addresses We will address the challenge by using machine learning and deep learning approaches to predict combinations of protein-protein interactions based on learning data. After that, this chapter also provides the briefest application of PPIs.

REFERENCES

[1] Rafael A Jordan, EL-Manzalawy Yasser, Drena Dobbs, Vasant Honavar. Predicting protein-protein interface residues using local surface structural similarity. *BMC bioinformatics*, 13, 1, (2012), 41.

[2] Ofran, Y; Rost, B. Analysing six types of protein-protein interfaces, *Journal of Molecular Biology*, vol. 325, no. 2, pp. 377–387, 2003. View at: Publisher Site|Google Scholar.

[3] Thomas, N Kipf, Max Welling. Semi-supervised classification with graph convolutional networks. arXiv preprint *arXiv:1609.02907*, (2016).

[4] Ruben Sanchez-Garcia, Carlos Oscar Sánchez Sorzano, José María Carazo, Joan Segura. BIPSPI: a method for the prediction of partner-specific protein– protein interfaces. *Bioinformatics*, 35, 3, (2019), 470–477.

[5] Mile Šikić, SanjaTomić, KristianVlahoviček, Prediction of protein–protein interaction sites in sequences and 3D structures by random forests. *PLoS computational biology*, 5, 1, (2009).

[6] Zhitao Ying, Jiaxuan You, Christopher Morris, Xiang Ren, Will Hamilton, Jure Leskovec. Hierarchical graph representation learning with differentiable pooling. In *Advances in neural information processing systems*., 2018, 4800–4810.

[7] Zaki, N; Lazarova-Molnar, S; El-Hajj, W; Campbell, P. "Protein protein based on pairwise similarity," *BMC bioinformatics*, vol. 10, no. 1, p. 150, 2009.

[8] Xenarios, I; Eisenberg, D. *Protein interaction databases*, *Current Opinion in Biotechnology*, vol. 12, no. 4, pp. 334–339, 2001.

[9] Kim, WK; Park, J; Suh, JK; et al., Large scale statistical prediction of protein-protein interaction by potentially interacting domain (pid) pair, *Genome Informatics Series*, pp. 42–50, 2002.

[10] Bader, GD; Hogue, CW. Analyzing yeast protein–protein interaction data obtained from different sources, *Nature biotechnology*, vol. 20, no. 10, pp. 991–997, 2002.

[11] Von Mering, C; Krause, R; Snel, B; Cornell, M; Oliver, SG; Fields, S; Bork, P. Comparative assessment of large-scale data sets of protein–protein interactions, *Nature*, vol. 417, no. 6887, pp. 399–403, 2002.

[12] Qi, Y; Bar-Joseph, Z; Klein-Seetharaman, J. Evaluation of different biological data and computational classification methods for use in protein interaction prediction, *Proteins: Structure, Function, and Bioinformatics*, vol. 63, no. 3, pp. 490–500, 2006.

[13] Xiuquan, D; Sun, S; Hu, C; Yao, Y; Yan, Y; Zhang, Y; Deep, PPI. Boosting prediction of protein protein interactions with deep neural networks, *Journal of Chemical Information and Modeling*, vol. 57, no. 6, 2017, pp. 1499 – 1510.

[14] Tanlin, S; Zhou, B; Lai, L; Pei, J. Sequence-based prediction of protein protein interaction using a deep-learning algorithm, *BMC Bioinformatics*, vol. 18, no. 1, 2017, pp. 277.

[15] Zhang, D; Kabuka, MR. Multimodal deep representation learning for protein-protein interaction networks, *2018 IEEE International Conference on Bioinformatics and Biomedicine (BIBM)*, pp. 595 – 602, 2018.

[16] Rumelhart, D; Hinton, G; Williams, R. Learning representations by backpropagating errors. *Nature*, 1986, 533–6.

[17] Goodfellow, I; Bengio, Y; Courville, *A. Deep learning*. MIT Press; 2016.

[18] Sotiris, B Kotsiantis, Zaharakis, I; Pintelas, P. Supervised machine learning: A review of classi_cationtechniques. *Emergingarti_cial intelligence applications in computer engineering*, 160, 3{24, 2007.

[19] Bernhard Scholkopf; Alexander J Smola. *Learning with kernels: support vector machines, regularization, optimization, and beyond*. MIT press, 2001.

[20] George, AF Seber; Alan J Lee. *Linear regression analysis*, volume 329. John Wiley & Sons, 2012.

[21] Anil K Jain. *Data clustering: 50 years beyond k-means. Pattern recognition letters*, 31(8), 651{666, 2010. doi: 10.1016/j.patrec.2009.09.011.

[22] Li Deng, Dong Yu; et al. Deep learning: methods and applications. *Foundations and Trends R in Signal Processing*, 7(3{4), 197{387, 201.

[23] Toshihiro Takahashi. *Statistical max pooling with deep learning*, July 3, 2018. US Patent 10,013,644.

[24] Volodymyr Mnih, Adria Puigdomenech Badia, Mehdi Mirza, Alex Graves, Timothy Lillicrap, Tim Harley, David Silver, Koray Kavukcuoglu. Asynchronous methods for deep reinforcement learning. In *International conference on machine learning*, pages 1928{1937, 2016.

[25] Jonathan Lorraine, David Duvenaud, Stochastic hyperparameter optimization through hyper networks. arXiv preprint *arXiv:1802.09419*, 2018.

[26] SergeyIo_e and Christian Szegedy. Batch normalization: Accelerating deep network training by reducing internal covariate shift. arXiv preprint *arXiv:1502.03167*, 2015.

[27] Abhishek Panigrahi, Yueru Chen, Jay Kuo, CC. Analysis on gradient propagation in batch normalized residual networks. arXiv preprint *arXiv:1812.00342*,2018.

[28] Toshihiro Takahashi. Statistical max pooling with deep learning, July 3 2018. US Patent 10,013,644.

[29] Alessandro Achille, Stefano Soatto. Information dropout: Learning optimal representations through noisy computation. *IEEE transactions on pattern analysis and machine intelligence*, 40(12), 2897{2905, 2018.

[30] Nathan Hubens. *Deep inside: Autoencoders - towards data science*, Apr 2018.

[31] Chaochun Liu, Yaliang Li, Hongliang Fei, Ping Li. Deep skip-gram networks for text classification. In *Proceedings of the 2019 SIAM International Conference on Data Mining*, pages 145{153. SIAM, 2019.

[32] Chaochun Liu, Yaliang Li, Hongliang Fei, Ping Li. Deep skip-gram networks for text classification. In *Proceedings of the 2019 SIAM*

International Conference on Data Mining, pages 145{153. SIAM, 2019.

[33] Zhou, Z. Learnware: on the future of machine learning. *Frontiers of Computer Science*, 2016, 10(4), 589–590.

[34] Rong, W; Peng, B; Ouyang, Y; Li, C; Xiong, Z. Structural information aware deep semi-supervised recurrent neural network for sentiment analysis. *Frontiers of Computer Science*, 2015, 9(2): 171–184.

[35] Sainath, TN; Vinyals, O; Senior, A; Sak, H. Convolutional, long short-term memory, fully connected deep neural networks. In: *Proceedings of IEEE International Conference on Acoustics, Speech and Signal Processing*., 2015, 4580–458.

[36] Dyer, C; Ballesteros, M; Ling, W; Matthews, A; Smith, NA. Transition based dependency parsing with stack long short-term memory. *Computer Science*, 2015, 37(2), 321–332.

[37] Sak, H; Senior, A; Beaufays, F. Long short-term memory based recurrent neural network architectures for large vocabulary speech recognition. In: *Proceedings of the 15 Annual Conference of the International Speech Communication Association*. 2014

[38] Li, Z; Wang, Y; Zhi, T; Chen, T. A survey of neural network accelerators. *Frontiers of Computer Science*, 2017, 11(5), 746–761.

[39] Lazib, L; Qin, B; Zhao, Y; Zhang, W; Liu, T. A syntactic path-based hybrid neural network for negation scope detection. *Frontiers of Computer Science*, 2020, 14(1), 84–94.

[40] Murakami, Yoichi. Network analysis and in silico prediction of protein–protein interactions with applications in drug discovery. *Current Opinion in Structural Biology*, vol. 44, pp. 134-142, 2017.

In: Deep Learning and its Applications
Editor: Arvind Kumar Tiwari
ISBN: 978-1-68507-185-1

Chapter 9

VARIOUS MACHINE LEARNING TECHNIQUES FOR SOFTWARE DEFECT PREDICTION

Sunil Malviya*
Department of Computer Science and Engineering,
Kamla Nehru Institute of Technology, Sultanpur, Uttar Pradesh, India

ABSTRACT

Today's majority of IT industries/companies demand automation. Software is the most important component for automation. Many companies or industries work for software development for specific or universal purposes. During the software development process, Software Defect Prediction (SoDP) is also an important term to use. In the active research areas of software engineering, SoDP plays a significant role. SoDP is finding the errors or faults or bugs or defects in software before deployment the software. For automation, the SoDP process is an important part of it. This chapter mainly focuses on various machine learning techniques, which are used for the prediction of defects in software. Most of the researchers are used supervised learning

* Corresponding Author's Email: sunilmalviya0104@gmail.com.

techniques to defect prediction in software. Similarly, many researchers also based on unsupervised learning techniques for the defect prediction in software. At present, the majority of researchers are focused on reinforcement learning techniques due to automation.

Keywords: software defect, defect prediction approaches, ML categories, evaluation performance

INTRODUCTION

Software Defect

A software or system defect is a fault, a bug, a mistake, or a shortcoming in a computer program or software that makes it return surprising or off-base outcomes. The existence of software bugs legitimately influences the quality and maintenance cost of software frameworks. Software flaws or defects are programming errors that result in different results than anticipated. The majority of the errors are caused by source code or design, but some are caused by incorrect code produced by compilers.

Software defects are a serious concern for both software developers and customers. Software defects not only reduce software quality and expense but also cause delays in the development process. To resolve this problem, software defect prediction (SoDP) is suggested. SoDP will effectively advance the efficacy of software testing and direct resource allocation. Software bugs/defects must be detected and fixed early in the SDLC to produce high-quality software.

To predict the software defect/bug event and to understand the process of software development life cycle (SDLC) are major parts or components that prompt achievement standards. Imperfection forecast utilizing Machine learning calculation could be implemented in any phase of SDLC, for example, recognizable feasibility study, analysis, structure, coding, verification, and sustainability, and any kind of SDLC model, for example, Iterative Model, Waterfall Model, Spiral Model,

Agile Model, V model, and Big Bang Model (Immaculate et al., 2019). Machine learning calculation and measurable investigation assist with speculating that a code is buggy.

In terms of Institute of Electrical and Electronics Engineers (IEEE) standard 104, software defects are classified in various ways (Park & Hong, 2014), which are:

Error: This can happen because the consumer is aware of the software's flaw, which causes incorrect performance.

Fault: There has been an apparent mistake in the software products.

Failure: Stop the software products from performing their previous necessary functions, or incorrect results will be stored for each customer feedback.

Defect: The inability to carry out the system and tasks requirements and specifications specified by the customers and developers.

Bug: A software bug is a mistake or deficiency in the source code of a program or software. It is detected by either developers or testing team members.

Types of Software Defects

1. Requirement defects: These kinds of defects occur when incorrect requirements are established. The best ways to detect such types of defects are by inspection. Testing can prove to be costly because developing a system on a bad set of requirements and then when it fails, having to re-develop it.
2. Design defects: These defects occur when the system is improperly designed. Numerous empirical studies have committed that analyses were much more accurate and efficient than testing. A flaw discovered during design inspection is reasonable to correct than one find out during feature testing because the cost of rework in the latter is significantly higher.

3. Code defects: For these types of defects, functional or structural testing is found to be better than inspection. According to some research, testing and inspection find various types of code defects and can thus be used in tandem to complement each other.

Software Defect Prediction

In the area of software engineering, software defect prediction (SoDP) is a significant research area. The method of identifying faulty modules in software is known as software defect prediction or SoDP. SoDP advantages to optimize testing resource distribution by identifying defect-prone modules aforementioned to testing. Until deploying software systems, most companies want to estimate the number of bugs or recognize defect-prone modules. When a software is exposed to operation or during research, a diversity of statistical approaches and artificial intelligence (AI) techniques have been employed for defects prediction.

The following important areas during software development activities for prediction of software defects:

- Predict the software defect in software systems
- Improve the software quality
- Decrease the maintenance cost
- Decrease the efforts

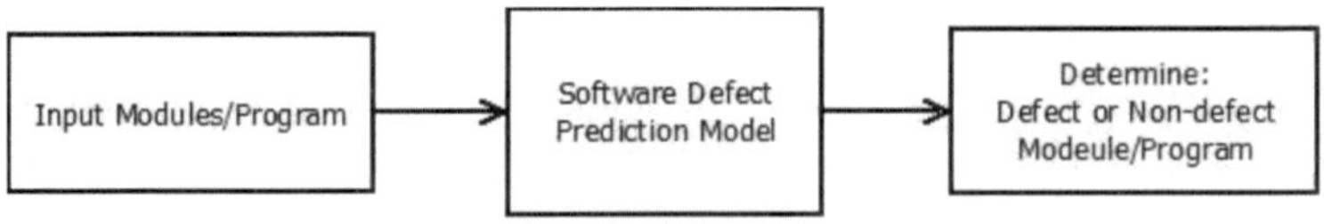

Figure 1. The basic structure of the SoDP model.

Software defect prediction is the most popular research area in software engineering. Software defect prediction works based on historical data. To construct machine learning classifiers to predict faulty code snippets, software defect prediction is a method of using historical data from software archives/repositories such as code complexity and change records to create software defect metrics.

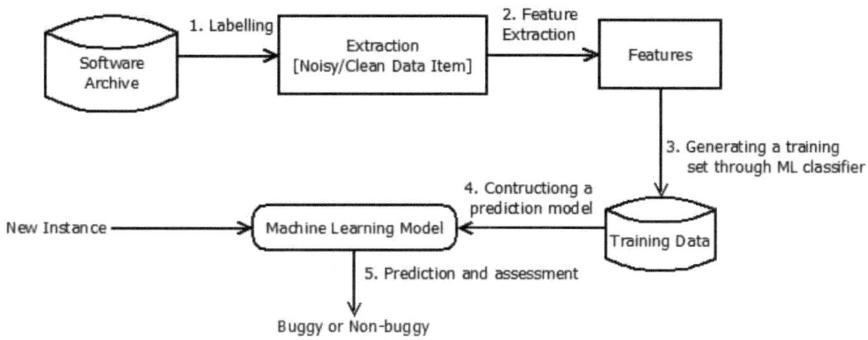

Figure 2. Process of software defect prediction model.

The following steps for the process of software defect prediction model (as seen in Figure 2) are:

- Step 1: The initial step in creating a prediction model is to gather data from software achieve such as e-mail collections, version control systems, and problem tracking systems.
- Step 2: In keeping with prediction granularity, each one instance may characterize a device, a source code file, software component, a function, a class, and a code change.
- Step 3: In this situation, many metrics (or features) are extracted from program repositories and are labeled as buggy/clean or the number of bugs.
- Step 4: Apply pre-processing techniques after producing instances with metrics and labels.
- Step 5: The final step is to train a prediction model that can guess whether or not a new instance contains a bug.

Brief History of Software Defect Prediction Studies

In the year 1971, the author (Akiyama) performed the first research to estimate the number of defects/faults. The author created a simple model using LOC (Lines of Code) to reflect the complexity of software systems, based on four assumptions that complex source code might reason bugs/defects. However, LOC is an overly simplistic metric for illustrating machine complexity (Nam, n.d.). In 1976 and 1977, MaCabe and Halstead introduced the cyclomatic complexity metric and the Halstead complexity metric, respectively. In the 1970s and early 1980s, these metrics were widely popular to develop models for appraising defects (Benton & Neil, 1999).

The models examined during that time were not prediction models, but rather fitting models that looked at the relationship between metrics and the number of defects (Benton & Neil, 1999). The authors developed a linear regression model and tested it on a different set of program codes (Shen et al., 1985). Munson et al., on the other hand, argued that at the time, advanced regression techniques were not defined, and suggested a classification model that divides modules into two classes: low risk and high risk (Munson & Khoshgoftaar, 1992).

Chidamber and Kemerer (Chidamber & Kemerer, 1994) suggested various object-oriented (OO) metrics in 1994, which were used by Basili et al., (Basili et al., 1996) to predict defects in object-oriented systems.

Defect/Bug Life Cycle

Predicting the existence of a defect and recognizing the life-cycle of defects are most important and compulsory, indicating the significance of predicting the defect/bug prior in the SDLC. Defect life cycle saves time, effort, and money in finding and repairing the defects during the software development. The number of defects/bugs posed throughout the software development cycle is one of the major issues confronting the modern software company/industry. This results in a delay in the product's final delivery and an overall rise in the running cost. The model attempts to forecast the existence of the bug by using parameters from different dimensions. Most well-known software development models, such as waterfall, agile, V-shape, and spiral, are well-suited to the model (Immaculate et al., 2019).

One of the most important thing is to comprehend the defect management life cycle. Figure 3 presents the various states of defect life cycle which are as follows:

1. New: A new defect is raised for the first time by testers or someone else a portion of the SDLC.
2. Assigned: Each new bug is allocated to any one of the team members for instant determination, depending on the bug's priority and seniority, and the equivalent developer will take action.
3. Open: If the preliminary state of the developer's testing reveals that it is indeed a bug, it will be moved to the open state. A developer team or a product testing team will usually handle this.
4. Fixed: All open bugs should be patched according to their severity and priority. Throughout the software development life cycle, all stakeholders collaborate to determine the importance and severity of bugs.
5. Retest: If the bug isn't checked or the explanation isn't accurate, it will require further testing to ensure that the reported bug is correct in all situations.
6. Reopened: The reopened bugs are important because the amount of reopened bugs increases the service time. It takes longer to patch reopened bugs.
7. Rejected: Rejected bugs aren't even bugs, but they were brought up by the testing team member. These are largely attributable to an ability set mismatch and specifications miscommunication.
8. Deferred: Some of the vulnerabilities will be pushed to the next updates due to priorities. Differentiated bugs are the classification for these bugs.

9. Duplicated: When the same bug occurs two or more times, we refer to it as a repeated bug.
10. Closed: This would be viewed as fixed bugs if the problem is correctly fixed.

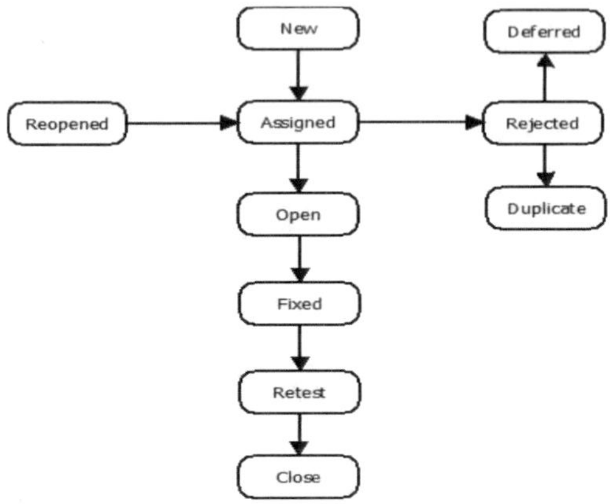

Figure 3. Defect/bug life cycle (Immaculate et al., 2019).

In this chapter, the introduction section discussed the process of software defect prediction in different phases/steps wise. Then we will explain the various ML categories (e.g., Supervised, Unsupervised or Semi-supervised Learning) to be used to predict the defect in a software. After that, will discuss various approaches (WPDP, CPDP, JITDP) are applied to software defect prediction. Then, we will move on to various measurement methods applied for predicting the software defects. And finally, we will conclude the chapter in the conclusion section.

Different Categories of Machine Learning

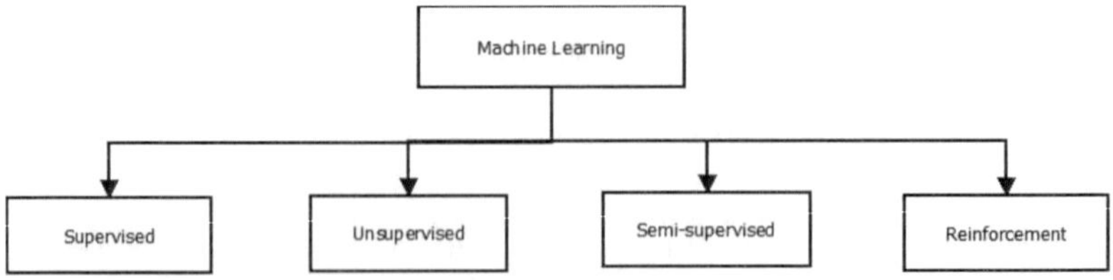

Figure 4. Different categories of Machine Learning.

Supervised Learning

Where a model must be trained to make a prediction, supervised learning is used. Regression and classification are two different methods of supervised learning (Lamba et al., 2019). To construct the prediction model, Machine learning classification (supervised) algorithms can be used with previous software metrics and fault marks. Concept learning, classification, rule learning, bayesian learning, LR, NN, and SVM are some of the most popular supervised machine learning approaches.

For bug prediction, Lamba et al., (2019) used a variety of machine learning algorithms, e.g., Linear regression, RF, NN, SVM, and DT. Standard defect prediction datasets were used to test the performance of these methods. In the term of bug prediction, the SVM approach outperformed all other methods.

To detect the software defects, Mori et al., (2019) proposed a new classification model which is called Superposed Naive Bayes (SNB). The different classification models along with SNB were simulated on the 13 real-world public datasets and found that SNB achieved a fine balance between accuracy and interpretability.

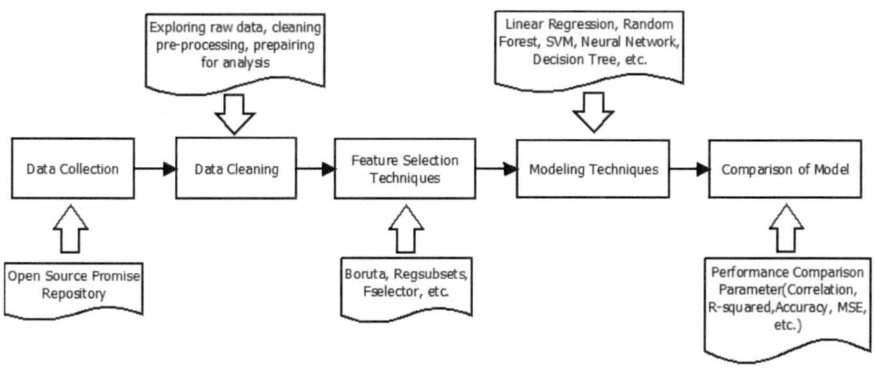

Figure 5. A framework for SoDP via previous databases (Lamba et al., 2019).

Immaculate et al., (2019) reported RF and performed better than various machine learning algorithms. Adequate priorities were not provided on the selection of variables, transformations of variables, variable reductions, and feature engineering that could enhance the efficiency of the models. Various algorithms are currently available which could be practiced to predict the bug's existence. Because of the ensemble nature, the random forest algorithm is the favorite option.

Shepperd et al., (2014) were using a novel benchmark method to predict and test software defects. During the assessment point, various learning schemes are assessed based on the system chosen. The most excellent learning scheme is then used to create a predictor with all historical/previous data in the prediction stage, and the analyst/developer is After that, it was used to predict a defect in new data.

Unsupervised Learning

When constructing unsupervised models using clustering algorithms, data sets have no class label (fault or non-fault) of each module. Unsupervised learning is a technique for detecting hidden patterns in input data. Clustering is a form of unsupervised learning (Lamba et al., 2019). Sequential pattern mining, association rule mining, and clustering are the most popular unsupervised learning methods.

Öztürk et al., (2015) investigated clustering methods performed in terms of defect prediction. Four variants of the K-mean clustering technique were considered in their study. Four real-world datasets have been used to evaluate the performance of different variants. According to the authors, the K-mean++ variant outperforms other K-mean variants.

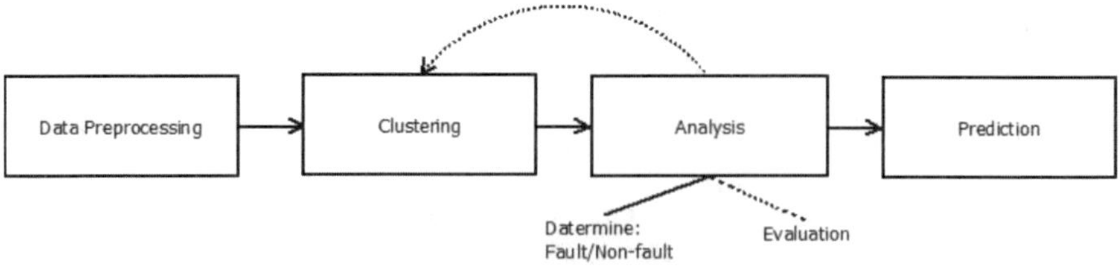

Figure 6. The process of constructing an unsupervised learning model.

Balogun et al., (2019) used several clustering techniques to solve software defect prediction challenges and compared their results. For defect prediction problems, K-mean, X-mean, hierarchal clustering, density-based clustering, and Expectation minimization methods were studied.

Due to the absence of historical data, Marjuni et al., (2019) have used an unsupervised approach to predict software defects. A signed Laplace spectral classifier was proposed to analyze defects. According to the simulation results, the suggested signed classifier outperformed the unsupervised technique significantly.

During the software development cycle, Yadav et al., (2015) developed a fuzzy-based approach for dealing with software defects. Twenty real-world datasets were used to test the suggested method. The proposed approach's accuracy was found to be close to the real defect prediction rate.

Singh et al., (2017) developed an artificial framework for extracting software defect fuzzy rules. The proposed model was capable to identify fault features. The model started with the assumption that every feature was a meaningless feature. The suggested framework's performance was evaluated using publicly available software defect datasets. It was shaw that the suggested model could find fuzzy rules for software flaws.

Semi-Supervised Learning

Semi-supervised learning combines the benefits of both supervised and unsupervised learning. Semi-supervised learning is in the middle of the two types of learning: supervised and unsupervised. As compared to unsupervised learning, semi-supervised clustering increases efficiency, but unsupervised learning does not perform as compared to supervised learning. Semi-supervised learning is a form of ML that trains models using a combination of a lesser quantity of labeled data and a huge quantity of unlabeled data.

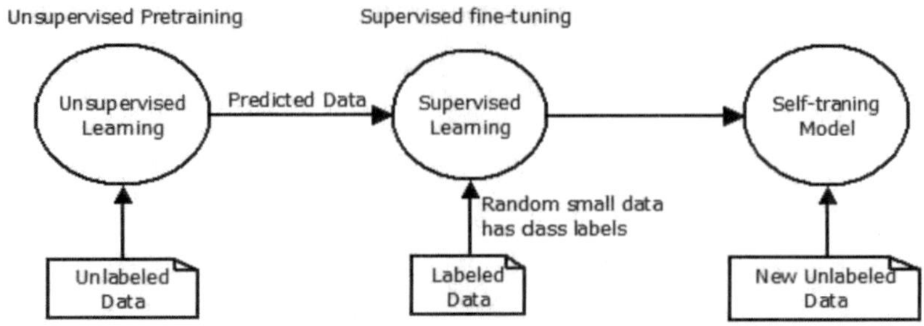

Figure 7. The proposed semi-supervised learning framework (Chen et al., 2020).

The software engineering community has attempted to incorporate different machine learning methods such as active or semi-supervised learning for defect prediction as they have been suggested. Lu et al., (Lu & Cukic, 2012) proposed defect prediction models based on active learning, in which a trial collection of occurrences is chosen and the occasions are queried by professionals.

Li et al., (2012) are suggested sample-based methods for predicting software defects. A smaller number of modules were chosen and tested for the large software framework. A software defect prediction model was created to forecast defect proneness in the modules on the left. Random sampling with a semi-supervised trainer, traditional computer learners, and active sampling with an active semi-supervised learner are the three sample selection methods mentioned in that paper.

A semi-supervised organized dictionary learning (SSDL) technique was used by Wu et al., (2018). If there is an absence of historic data for constructing a reliable model, this method yields excellent results. They have derived with cross-project defect prediction and semi-supervised defect prediction (SSDP) as potential successful solutions. For cross-software defect prediction and within-software defect prediction, the authors suggested an SSDL approach. They combined a minor quantity of labeled defect data with a major quantity of unlabeled data. SSDL performs better than related SSDP approaches, according to their findings. In the CSDP case, this occurs for two datasets.

Semi-supervised learning's main objectives are to build a beginner that automatically unnecessary avails the large quantity of unlabeled data in addition to the small quantity of labeled data to help enhance learning results, as according to Chapelle et al., (2009). Low-density separation-based methods, Generative-model based methods, disagreement-based methods, and graph-based methods are the most popular semi-supervised learning methods.

Reinforcement Learning

Semi-supervised learning is not the same as reinforcement learning. Reinforcement learning is a system in which reward values are associated with the various actions that the model is expected to take. Reinforcement learning includes a set of actions, parameters, and end values. It will train the machine by trial and error method. It will learn from past efforts to achieve the best possible result.

In Figure 7, the authors are described how interact an agent with an environment? How to improve or enhance the states? How to define the set of actions? Let us consider it as the state with S, action with A, and reward with R If. An agent completes an action and reaches the environment so that Ai will increase by 1, where i means 1, 2, 3, and so on. Similarly, the environment will give an acknowledgment in terms of reward as well as a state will increase with 1, so that set of the state is Si. As seen in Figure 7, this results in a series of states Si, action Ai, and immediate rewards Ri. The agent's task is to figure out a control policy, n: S + A, that maximizes the expected amount of these rewards, with future rewards discounted exponentially by their delay (Mitchell, 1997).

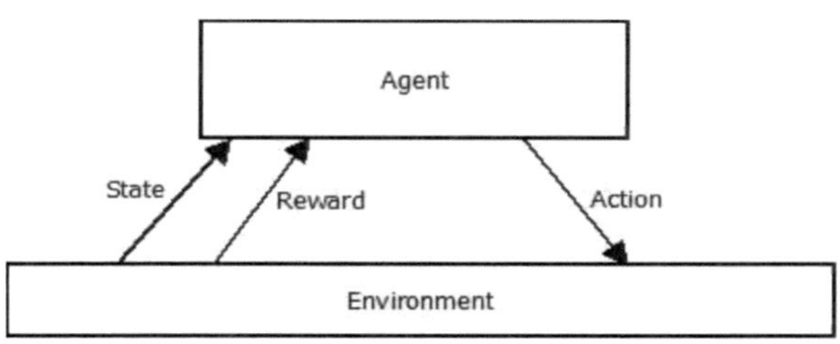

Figure 8. A framework of Reinforcement Learning (Mitchell, 1997).

SOFTWARE DEFECT PREDICTION APPROACHES

Within-Project Defect Prediction

The machine learning techniques used, as well as the consistency of the training data, influence a model's predictive power. In most cases, defect prediction models are studied in the literature in a within-project sense, assuming that previous defect data for a specific project exists (Herbold et al., 2017). Within-project defect prediction (WPDP) is the name of this process. WPDP is a prediction model that can be built by collecting historical data from a software project and predicts faults in that project. If there is a satisfactory quantity of historical data to train models, WPDP performed best.

Many new technologies and development methods will impact the new code submitted after the upgrade, and could even introduce new defects in software development, according to the within-project defect prediction. As a result, begin by sorting the data by time series, then divide the data from previous years into a category, and choose the data from the previous year as the training set and the data from the following year as the test set. That does not only ensure the integrity of a project development period but also that the amount of data in the training set for a few years is adequate to improve the accuracy of the with-in project defect prediction.

Cross-Project Defect Prediction

Cross-project defect prediction (CPDP) is the process of forecasting defects in other projects using prediction models trained on data from previous projects. When a project does not have enough historical data to train a model, CPDP is used in this mode. As a result, a prediction model is created for one project and then applied to another or several projects. The prediction models have been transferred from one project to another project.

The effect of a predictor in machine learning-based predictions is determined by 2 aspects: the learning algorithm and training data. As a result, there are two possible approaches to CPDP (Zimmermann et al., 2009; He et al., 2012):

1. Identifying the most appropriate training data for the project that needs to be predicted. In an ideal world, we'd be able to find learning data with a similar flaws pattern as the goal project, resulting in satisfactory prediction performance.
2. To build the defect predictor, we used learning algorithms with a high generalization potential. This approach assumes that is a consistent defect pattern across all datasets. If it can effectively train this pattern, it can foresee flaws in another project.

Just-in-Time Defect Prediction

JITDP (Just-in-Time Defect Prediction) is a form of SoDP approach that allows software change-level predictions. It detects defect-causing program changes as soon as made (i.e., "just-in-time"). Inspection is much simpler at this point because the improvements are still new in the developers' minds (Cabral et al., 2019).

Advantages of JITDP over component level SoDP include (Kamei et al., 2013):

1. Prediction is made early in the process, allowing for easier code inspection.
2. Predictions with a finer granularity, making it easier to see defects.
3. Developers are assigned to inspect the code straightforwardly.
4. Debugging code after a defect report has been produced.
5. SoDP, which focuses on predicting flaws in software components (e.g., files, packages, etc.), is a more traditional approach (Cabral et al., 2019).

PERFORMANCE EVALUATION OF SODP

In the circumstance of SoDP, defect-prone classes are generally measured as positive occurrences/instances (Jiang et al., 2008; He et al., 2012).

	Actual Postive	Actual Negative
Predicted Positive	TP	FP
Predicted Negative	FN	TN

Figure 9. The confusion matrix (Khan et al., 2019).

As a consequence, the following four classifications of defect prediction results are:

1. True-Positive (TP): defect-prone groups that are grouped correctly
2. False-Negative (FN): defect-prone groups that are incorrectly grouped to be defect-free
3. True-Negative (TN): defect-free groups that are grouped correctly
4. False-Positive (FP): the defect-free group that are incorrectly grouped to defect prone

Some of the classification evaluation measurements are:

False Positive Rate

The optimum value for FPR is 0. False Positive Rate (FPR) is the proportion of instances classified as non-defective that were wrongly predicted as defective.

$$pf = FPR = \frac{FP}{TN + FP} \quad (1)$$

Accuracy

The percentage of correctly categorized elements is known as accuracy (both as defect-prone and not defect-prone). 1.0 is the best accuracy value.

$$\text{Accuracy} = \frac{\text{TP} + \text{TN}}{\text{TP} + \text{FP} + \text{TN} + \text{FN}} \tag{2}$$

Precision

Precision refers to how much of the elements reverted by a model were defect-prone in the first place. The precision value of 1.0 is the best; the higher the precision, the fewer false positives (i.e., elements incorrectly classified as defect-prone).

$$\text{Precision} = \frac{\text{TP}}{\text{TP} + \text{FP}} \tag{3}$$

Recall/True Positive Rate

The number of defect-prone elements currently returned by a model is referred to as recall. The highest recall value is 1.0; the higher the recall, the lower the number of false negatives (i.e., elements missed by the model).

$$\text{pd} = \text{Recall} = \frac{\text{TP}}{\text{TP} + \text{FN}} \tag{4}$$

F-Measure/Score

The F-score is often referred to as the F1-score. It's a metric for how accurate a model is on a given dataset. It's used to assess binary classification systems that categorize examples as either "positive" or "negative." The F-score, which is known as the harmonic mean of the model's precision and recall, is a way of combining the model's precision and recall.

$$\text{F} - \text{measure} = \frac{2 * \text{Precision} * \text{Recall}}{\text{Precision} + \text{Recall}} \tag{5}$$

Area under the Curve (AUC)

An outstanding model has an AUC close to 1, indicating that it has a high level of separability. AUC close 0 indicates a weak model, which means it has the lowest measure of separability.

Receiver Operating Characteristic (ROC)

A Receiver Operating Characteristic curve, also known as the ROC curve. It is a graphical demonstration of the binary classifier system's true predictive power when its threshold is wide-ranging discriminately. At different threshold settings, the ROC curve plots the true positive rate on the x-axis and the false positive rate on the y-axis. The AUC-ROC curve is an output measurement for classification problems at different threshold settings. AUC represents the degree or metric of separability, and ROC is a probability curve (Understanding AUC - ROC Curve | by Sarang Narkhede | Towards Data Science, n.d.).

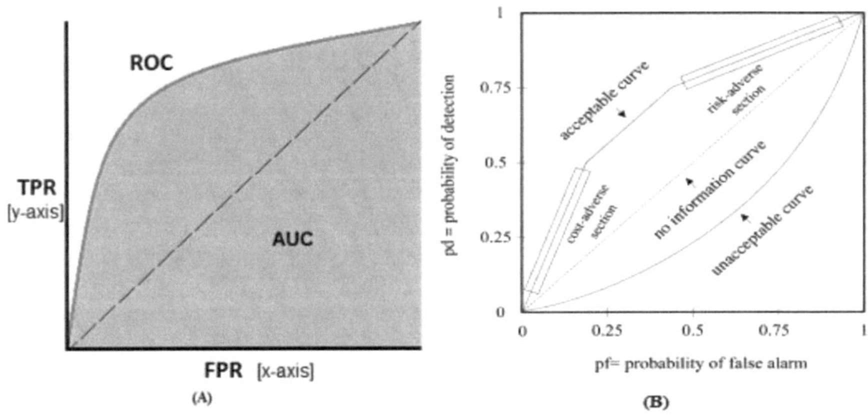

Figure 10. (A) AUC - ROC curve (B) AUC-PR [ROC curves] (Arar & Ayan, 2015).

In a ROC curve, balance is equal to the normalised Euclidean distance from the anticipated point (0, 1) to (pf, pd). as following:

$$\text{balance} = 1 - \frac{\sqrt{(1-\text{pd})^2 + (0-\text{pf})^2}}{\sqrt{2}} \quad (6)$$

where ROC means Receiver Operating Characteristic, AUC means Area Under the Curve, FPR means False Positive Rates at x-axis (pf), TPR means True Positive Rates at y-axis (pd), PR means Precision and Recall.

The gray distribution curve in the diagrams below is for the positive class, while the black scattering curve is for the negative class. The following cases are considering for the AUC-ROC curve.

Case 1

If the AUC value is 1 means True Negative (TN) and True Positive (TP) both are not overlapped to each other. This case is the best case for all time. In this case, the defects are 100% sure to be recognizable.

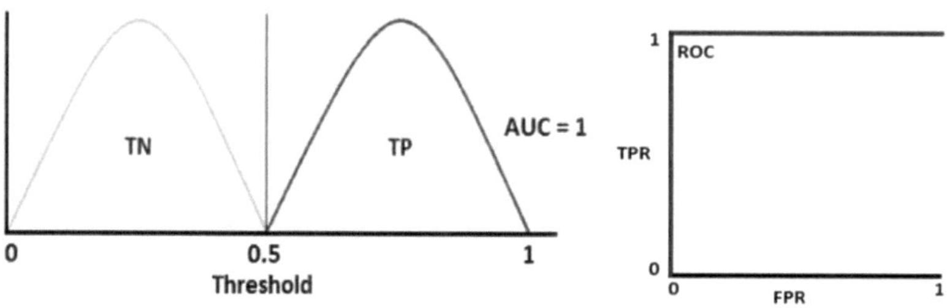

Case 2

If the AUC value is 0.7 means True Negative (TN) and True Positive (TP) both are overlapped to each other. In this case, the defects are 70% sure to be recognizable.

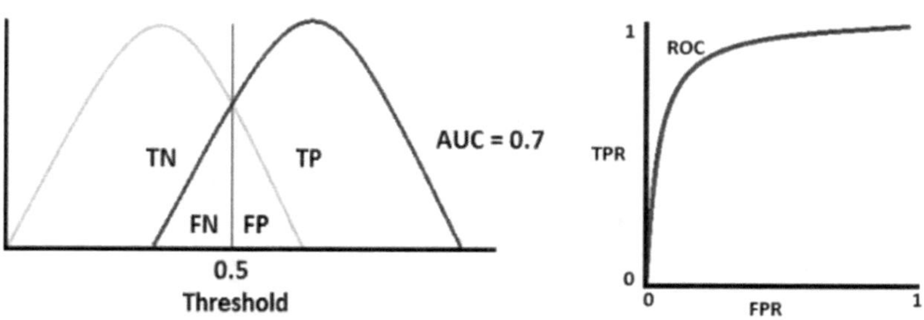

Case 3

If the AUC value is 0.5 means True Negative (TN) and True Positive (TP) both are equally overlapped to each other. This case is an average case. In this case, the defects are 50% sure to be recognizable.

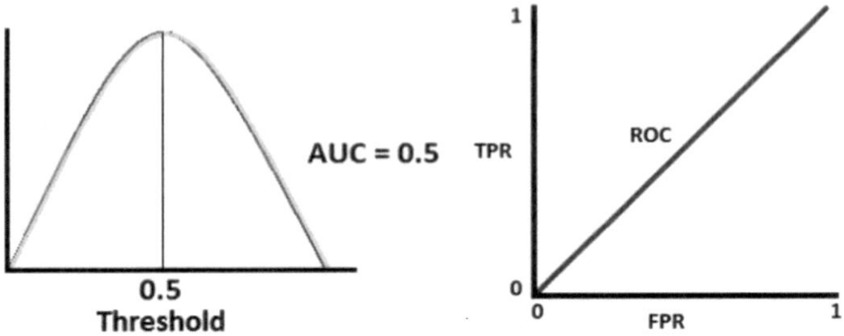

Case 4

If the AUC value is 0 means True Negative (TN) and True Positive (TP) both are not overlapped to each other, but both classes are interchangeable. This case is the worst case of all time.

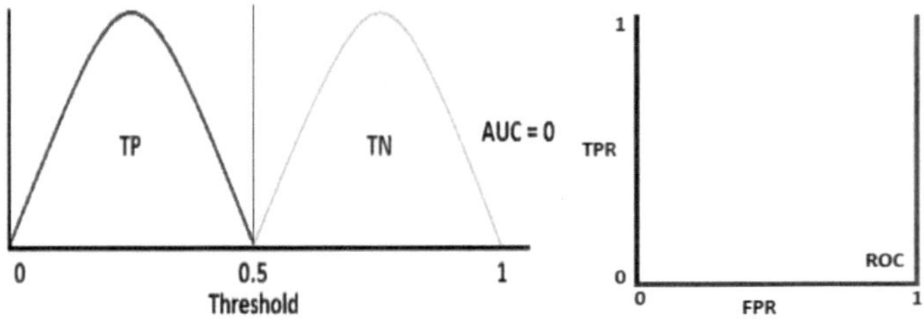

CONCLUSION

In the active research areas of software engineering, Software Defect Prediction (SoDP) plays an important role in detecting software defects. This chapter helps to understand various software defects and SoDP model. Also a brief histroy of software defect prediction along with the defect life cycle of software is discussed. Due to the popularity of Machine learning (ML) techniques in field of software engineering they are used to predict software defects. Various ML categories such as Supervised, Unsupervised, Semi-supervised, and Reinforcement learning are applied on different defect prediction approaches such as with-in project, cross-project, and just-in-time project. And in last, different performance evaluation measures to predict the software defects are elaborated.

REFERENCES

Arar, Ö. F., & Ayan, K. (2015). Software defect prediction using cost-sensitive neural network. *Applied Soft Computing Journal*, 33, 263–277. https://doi.org/10.1016/j.asoc.2015.04.045

Balogun, A. O., Oladele, R. O., Mojeed, H. A., Amin-Balogun, B., Adeyemo, V. E., & Aro, T. O. (2019). Performance Analysis of Selected Clustering Techniques for Software Defects Prediction Aro. © *Afr. J. Comp. & ICT*, 12 (2), 30–42. https://afrjcict.net.

Benton, N. E., & Neil, M. (1999). A critique of software defect prediction models. *IEEE Transactions on Software Engineering*, 25 (5), 675–689. https://doi.org/10.1109/32.815326.

Cabral, G. G., Minku, L. L., Shihab, E., & Mujahid, S. (2019). Class Imbalance Evolution and Verification Latency in Just-in-Time Software Defect Prediction. *2019 IEEE/ACM 41st International Conference on Software Engineering (ICSE)*, 2019-May, 666–676. https://doi.org/10.1109/ICSE.2019.00076.

Chen, T., Kornblith, S., Swersky, K., Norouzi, M., & Hinton, G. (2020). *Big Self-Supervised Models are Strong Semi-Supervised Learners.* ArXiv. http://arxiv.org/abs/2006.10029.

He, Z., Shu, F., Yang, Y., Li, M., & Wang, Q. (2012). An investigation on the feasibility of cross-project defect prediction. *Automated Software Engineering*, 19 (2), 167–199. https://doi.org/10.1007/s10515-011-0090-3.

Herbold, S., Trautsch, A., & Grabowski, J. (2017). Global vs. local models for cross-project defect prediction: A replication study.

Empirical Software Engineering, 22 (4), 1866–1902. https://doi.org/10.1007/ s10664-016-9468-y.

Immaculate, S. D., Begam, M. F., & Floramary, M. (2019, March 1). Software Bug Prediction Using Supervised Machine Learning Algorithms. *2019 International Conference on Data Science and Communication, IconDSC 2019.* https://doi.org/10.1109/IconDSC.2019.8816965.

Jiang, Y., Cukic, B., & Ma, Y. (2008). Techniques for evaluating fault prediction models. *Empirical Software Engineering*, 13 (5), 561–595. https://doi.org/10.1007/s10664-008-9079-3.

Kamei, Y., Shihab, E., Adams, B., Hassan, A. E., Mockus, A., Sinha, A., & Ubayashi, N. (2013). A large-scale empirical study of just-in-time quality assurance. *IEEE Transactions on Software Engineering*, 39 (6), 757–773. https://doi.org/10.1109/TSE.2012.70.

Khan, Shahzad Ali, Rana, Z. A. (2019). Evaluating Performance of Software Defect Prediction Models Using Area Under Precision-Recall Curve (AUC-PR). *2nd International Conference on Advancements in Computational Sciences (ICACS). IEEE*, 2019, 1–6. https://doi.org/10. 23919/ICACS.2019.8689135.

Lamba, T., Kavita, & Mishra, A. K. (2019). Optimal Machine learning Model for Software Defect Prediction. *International Journal of Intelligent Systems and Applications*, 11 (2), 36–48. https://doi.org/10. 5815/ijisa.2019.02.05.

Marjuni, A., Adji, T. B., & Ferdiana, R. (2019). Unsupervised software defect prediction using signed Laplacian-based spectral classifier. *Soft Computing*, 23 (24), 13679–13690. https://doi.org/10.1007/s00500-019-03907-6.

Mitchell, T. M. (1997). Machine Learning. In McGraw-Hill Science/Engineering/Math. *McGraw-Hill Science/Engineering/Math*. http://www.cs.cmu.edu/afs/cs.cmu.edu/user/mitchell/ftp/mlbook.html

Mori, T., & Uchihira, N. (2019). Balancing the trade-off between accuracy and interpretability in software defect prediction. *Empirical Software Engineering*, 24 (2), 779–825. https://doi.org/10.1007/s10664-018-9638-1.

Nam, J. (2014). *Survey on Software Defect Prediction.*

Öztürk, M. M., Cavusoglu, U., & Zengin, A. (2015). A novel defect prediction method for web pages using k-means++. *Expert Systems with Applications*, 42 (19), 6496–6506. https://doi.org/10.1016/j.eswa. 2015.03.013.

Park, M., & Hong, E. (2014). Software Fault Prediction Model using Clustering Algorithms Determining the Number of Clusters Automatically. *International Journal of Software Engineering and Its Applications*, 8 (7), 199–204. https://doi.org/10.14257/ijseia.2014. 8.7,16.

Shepperd, M., Bowes, D., & Hall, T. (2014). Researcher bias: The use of machine learning in software defect prediction. *IEEE Transactions on Software Engineering*, 40 (6), 603–616. https://doi.org/10.1109/TSE.2014.2322358.

Singh, P., Pal, N. R., Verma, S., & Vyas, O. P. (2017). Fuzzy Rule-Based Approach for Software Fault Prediction. *IEEE Transactions on Systems, Man, and Cybernetics: Systems*, 47 (5), 826–837. https://doi. org/10.1109/TSMC.2016.2521840.

Understanding AUC - ROC Curve | by Sarang Narkhede | Towards Data Science from https://towardsdatascience.com/understanding-auc-roc-curve-68b2303cc9c5.

Yadav, H. B., & Yadav, D. K. (2015). A fuzzy logic based approach for phase-wise software defects prediction using software metrics. *Information and Software Technology*, 63, 44–57. https://doi.org/10. 1016/j.infsof.2015.03.001.

Zimmermann, T., Nagappan, N., Gall, H., Giger, E., & Murphy, B. (2009). Cross-project defect prediction: A large scale experiment on data vs. domain vs. process. *ESEC-FSE'09 - Proceedings of the Joint 12th European Software Engineering Conference and 17th ACM SIGSOFT Symposium on the Foundations of Software Engineering*, 91–100. https://doi.org/10.1145/1595696.1595713.

ABOUT THE EDITOR

Dr. Arvind Kumar Tiwari is an Associate Professor in the Department of Computer Science and Engineering at KNIT, Sultanpur, UP, India. He received the BE degree in Computer Science and Engineering from CCS university, Meerut, M.Tech. in Computer Science and Engineering from UPTU, Lucknow and PhD in Computer Science and Engineering from IIT (BHU), Varanasi, India. He has more than 15 years of experience to guide UG and PG engineering students. He has more than 45 publications in journals of international repute including national journals, conferences and proceedings of the international conferences. He has also Published a book Ubiquitous Machine Learning and its Applications by IGI Global international publisher. He is a professional member of various societies like ACM, Internet Society of India ISTE etc. He has an experience to organise an International Conference and Faculty Development programs at KNIT Sultanpur. Presently he is guiding 05 PhD. Research Scholars. His research interests include Machine Learning, Computational Biology, Image Processing and Pattern Recognition.

Index

D

E

M

N

O

P

R

S

T